THE LITTLE BOOK OF NUMBER CHAINS

Dr Gareth Moore is the author of a wide range of puzzle and brain-training books for both children and adults, including *The Little Book of Sudoku* Volumes 4 and 5, *The Little Book of Crosswords*, *The Little Book of Cryptic Crosswords*, *The Kids' Book of Crosswords*, *The Little Book of Word Searches 1 & 2*, *The Kids' Book of Word Searches*, *The Brain Workout*, *Train the Brain* and *The 10-Minute Brain Workout*.

He writes the monthly logic puzzle magazine *Sudoku Xtra*, as well as the popular online puzzle site PuzzleMix.com. He gained his Ph.D at the University of Cambridge, UK, where he taught machines to recognize the English language.

THE LITTLE BOOK OF NUMBER CHAINS

Michael O'Mara Books Limited

First published in Great Britain in 2011 by
Michael O'Mara Books Limited
9 Lion Yard
Tremadoc Road
London SW4 7NQ

A CIP catalogue record for this book is available
from the British Library.

Papers used by Michael O'Mara Books Limited are natural, recyclable products made from wood grown in sustainable forests. The manufacturing processes conform to the environmental regulations of the country of origin.

ISBN: 978-1-84317-872-9

1 2 3 4 5 6 7 8 9 10

www.mombooks.com

www.drgarethmoore.com

Cover design by Ana Bjezancevic

Designed and typeset by Gareth Moore

Printed and bound in Great Britain by
CPI Cox & Wyman, Reading, RG1 8EX

Introduction

Good number skills are important for many reasons, from estimating the cost of shopping to understanding interest rates and planning your savings. Unfortunately nowadays many people rarely perform calculations in their heads, leaving us in the slow lane when it comes to mental arithmetic.

Just a few minutes of brain-bound number manipulation a day can make a real difference, and it's even claimed that improving your mental mathematical talents can lead to smarter thinking in seemingly unrelated areas.

In your hands you hold part of the solution: over 400 Number Chain workouts, as popularised in various daily newspapers, and all guaranteed to give your grey matter calculator a thorough challenge and maybe even a good spring clean too.

Each Number Chain consists of a series of six calculations to perform in sequence on an initial given number. Perform each numeric operation in order, from top to bottom as shown by the arrows, and write the result in the handy 'RESULT' box at the bottom.

Try if you can to do each chain of calculations 'in your head' without resorting to writing anything down. There's a mix of figures and words in the tasks - this forces your brain to work harder as you convert the words into numbers, and helps keep the puzzles fresh.

As you work through the book you'll find that the puzzles get harder as you go, and towards the end you'll find some much trickier mental manipulations to perform. Fully worked solutions are given at the back if you want to check your calculations.

Dr Gareth Moore (www.DrGarethMoore.com)

Puzzle 1

16

÷2

-5

+18

1/3 of this

Multiply by eleven

Subtract twelve

RESULT

Puzzle 2

11

-5

÷2

+35

×2

÷4

-15

RESULT

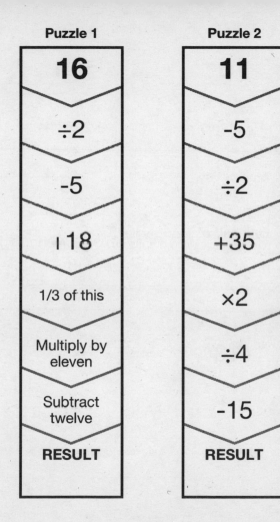

Puzzle 3

15
1/3 of this
Multiply by ten
One half of this
-2
+13
÷3
RESULT

Puzzle 4

40
1/2 of this
-6
÷7
×6
+5
-1
RESULT

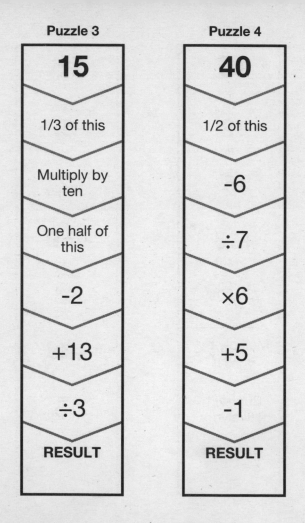

Puzzle 5

47

-11

√

Multiply by five

One third of this

+44

-25

RESULT

Puzzle 6

18

1/2 of this

Square root of this

+30

÷3

+16

Subtract twenty-two

RESULT

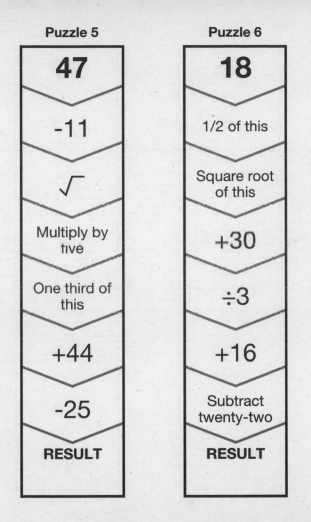

Puzzle 7

49

$\sqrt{}$

Add nineteen

50% of this

-2

+26

Subtract eight

RESULT

Puzzle 8

50

+27

-42

Divide by five

×4

+50%

1/3 of this

RESULT

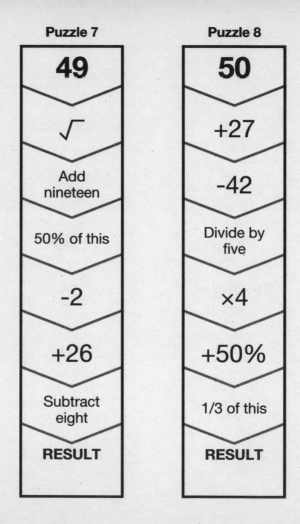

Puzzle 9

45

÷3

1/3 of this

Add fifty

Divide by five

-5

×10

RESULT

Puzzle 10

21

1/3 of this

Multiply by seven

+23

÷3

+75%

+5

RESULT

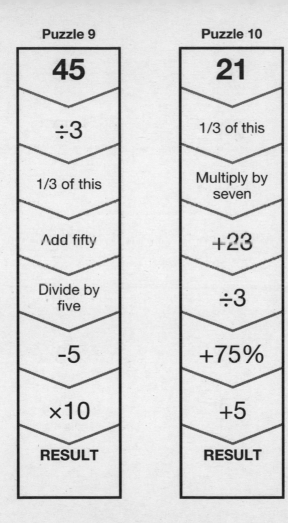

Puzzle 11

7

×4

1/2 of this

+1

÷3

+33

÷2

RESULT

Puzzle 12

39

Subtract four

÷7

+20

Subtract two

+3

Fifty percent of this

RESULT

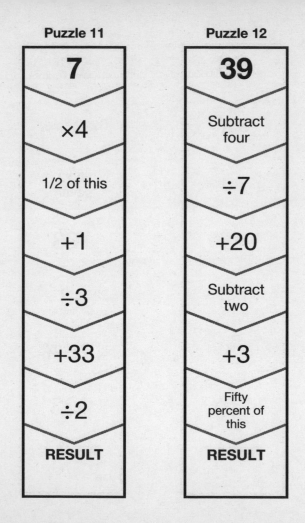

Puzzle 13

28
Fifty percent of this
-3
+1
÷3
+32
-1
RESULT

Puzzle 14

7
×5
÷7
+7
50% of this
1/2 of this
+33
RESULT

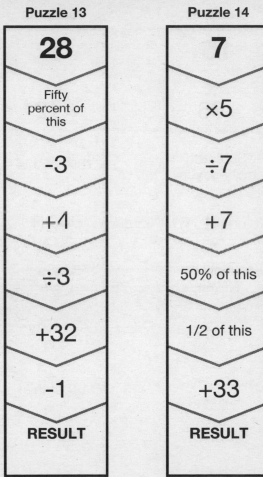

Puzzle 15

15
+26
Subtract thirty-seven
÷2
Multiply by eleven
1/2 of this
Add thirty-nine
RESULT

Puzzle 16

5
×12
75% of this
+1
-17
Add forty-one
1/2 of this
RESULT

Puzzle 17

17

+3

÷2

Fifty percent of this

+22

One third of this

√

RESULT

Puzzle 18

7

×8

One half of this

-4

÷12

Multiply by two

Add thirty-eight

RESULT

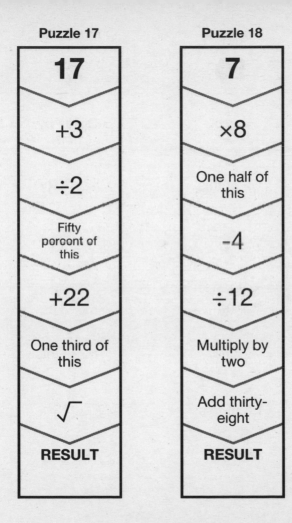

Puzzle 19

50
÷5
1/2 of this
×9
Subtract nine
+21
1/3 of this
RESULT

Puzzle 20

27
Divide by nine
+38
-39
×7
1/2 of this
+37
RESULT

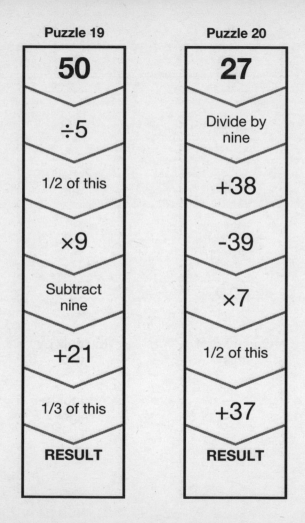

Puzzle 21

46

÷2

+18

Subtract thirty-one

50% of this

+46

÷3

RESULT

Puzzle 22

36

Multiply by two

25% of this

÷9

Add fifty percent

Add fourteen

-6

RESULT

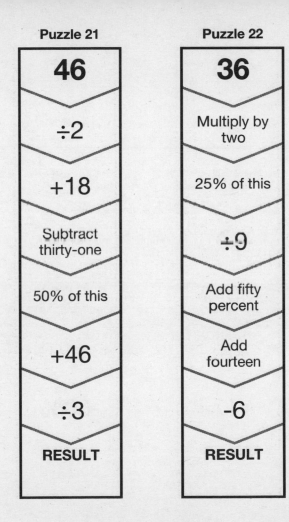

Puzzle 23

32
÷8
One half of this
×7
-50%
×8
+17
RESULT

Puzzle 24

13
-1
Divide by two
1/3 of this
+50%
+19
-50%
RESULT

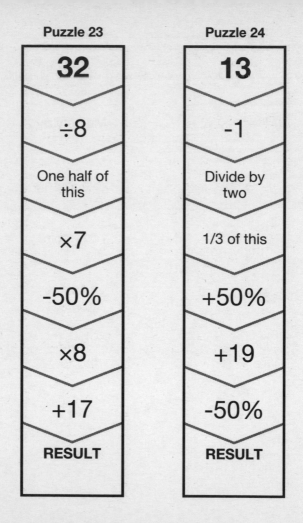

Puzzle 25

21
One third of this
×8
50% of this
÷7
+26
-50%
RESULT

Puzzle 26

41
-30
+39
1/2 of this
Add nineteen
÷11
×7
RESULT

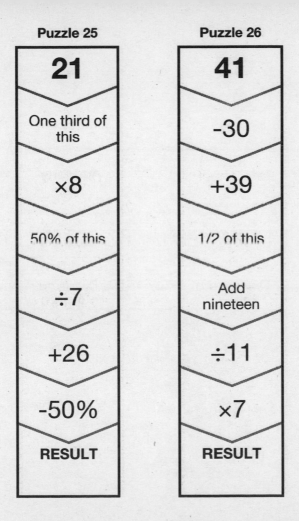

Puzzle 27

29
-8
×2
Add nine
One third of this
-13
×6
RESULT

Puzzle 28

6
1/2 of this
Add forty-nine
-25%
Divide by three
+29
50% of this
RESULT

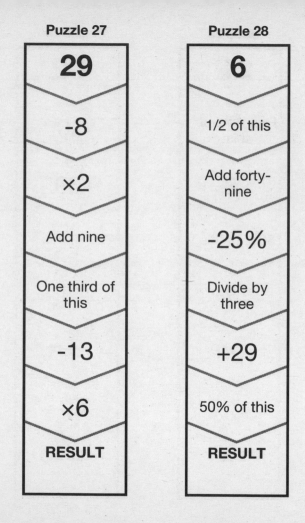

Puzzle 29

42
Divide by three
Multiply by two
+75%
Divide by seven
+36
-33
RESULT

Puzzle 30

32
Subtract eighteen
+24
One half of this
+40
-47
Twenty-five percent of this
RESULT

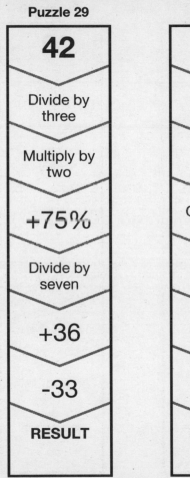

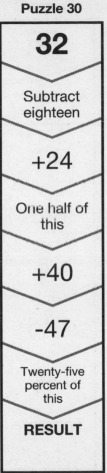

Puzzle 31

14
One half of this
×7
+23
÷6
1/4 of this
+15
RESULT

Puzzle 32

38
50% of this
+44
÷7
+36
1/3 of this
÷5
RESULT

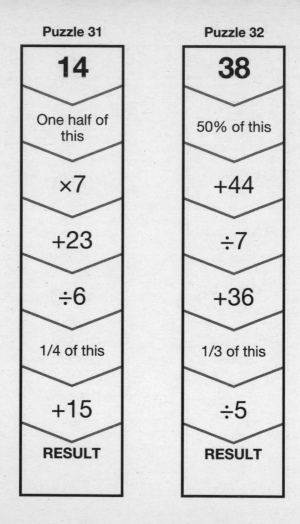

Puzzle 33

21
÷7
+22
Square root of this
Multiply by four
Divide by five
One half of this
RESULT

Puzzle 34

19
+45
1/4 of this
÷2
25% of this
Add fifty percent
×9
RESULT

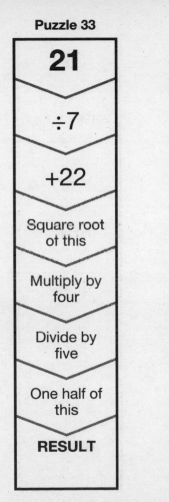

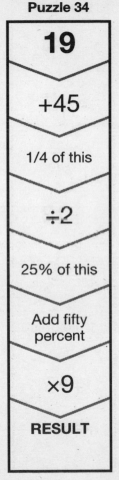

Puzzle 35

12
÷2
Add forty-seven
-13
+8
-4
1/4 of this
RESULT

Puzzle 36

34
50% of this
Add twenty-one
1/2 of this
+22
-11
÷2
RESULT

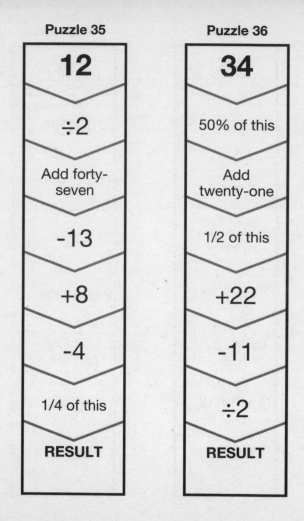

Puzzle 37

5

×10

-27

+33

-25%

1/3 of this

50% of this

RESULT

Puzzle 38

5

+38

-1

50% of this

-14

+3

Multiply by four

RESULT

Puzzle 39

18
+50%
One third of this
+11
1/4 of this
+29
-21
RESULT

Puzzle 40

36
Divide by two
×4
75% of this
One half of this
Subtract seven
÷4
RESULT

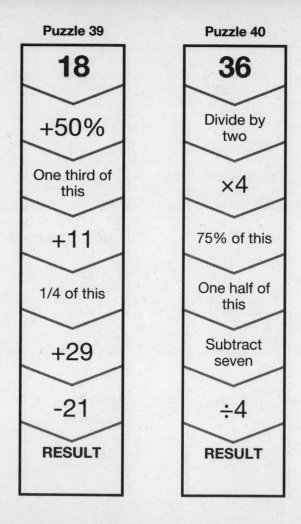

Puzzle 41

17
-7
One half of this
Multiply by five
-9
÷2
+47
RESULT

Puzzle 42

22
+18
1/2 of this
25% of this
Add forty-two
-14
÷3
RESULT

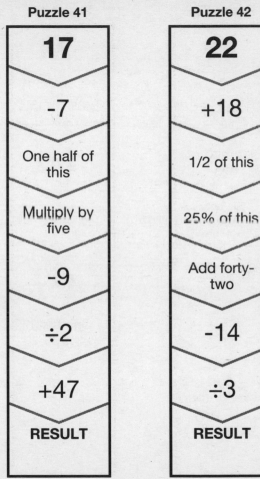

Puzzle 43

19
-15
×8
Seventy-five percent of this
One half of this
×4
Add twenty-four
RESULT

Puzzle 44

18
50% of this
+43
÷4
+19
One half of this
÷2
RESULT

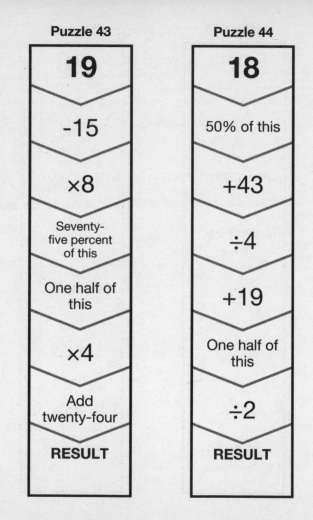

Puzzle 45

33
÷11
×7
1/3 of this
×8
Subtract thirty-seven
+42
RESULT

Puzzle 46

28
1/4 of this
+20
One third of this
$\sqrt{\ }$
+13
÷8
RESULT

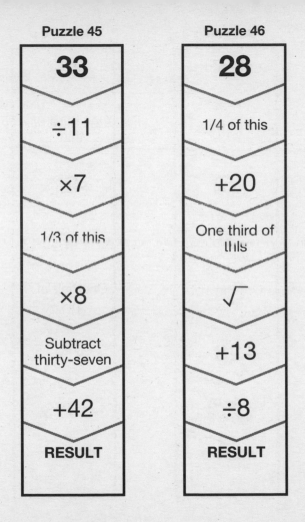

Puzzle 47

15
1/3 of this
×5
-3
÷2
+28
Divide by three
RESULT

Puzzle 48

21
One third of this
+21
÷7
One half of this
×5
-50%
RESULT

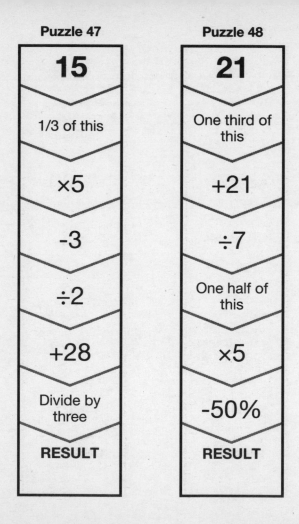

Puzzle 49

9

- √
- Add thirty
- 1/3 of this
- -6
- ×5
- +48

RESULT

Puzzle 50

37

- Subtract thirteen
- 1/3 of this
- ÷4
- +28
- ÷10
- +45

RESULT

Puzzle 51

28
+42
÷7
×4
+3
-21
One half of this
RESULT

Puzzle 52

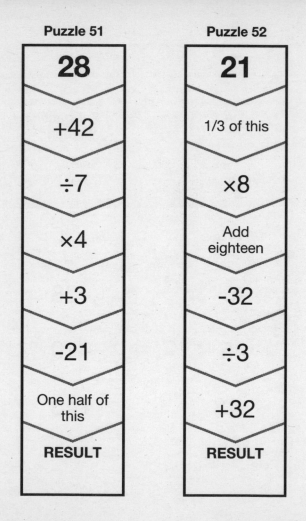

21
1/3 of this
×8
Add eighteen
-32
÷3
+32
RESULT

Puzzle 53	Puzzle 54
37	**12**
-33	÷6
+4	×5
×10	÷2
-26	×4
1/3 of this	+75%
-50%	Add seven
RESULT	**RESULT**

Puzzle 55

24
+25%
÷6
×2
+1
×7
-5
RESULT

Puzzle 56

34
Subtract fifty percent
×3
-25
One half of this
×5
+5
RESULT

Puzzle 57

38
1/2 of this
Multiply by three
+15
÷8
$\sqrt{\ }$
×2
RESULT

Puzzle 58

40
One half of this
50% of this
-5
+22
1/3 of this
Multiply by four
RESULT

Puzzle 59

26
÷2
+48
-20
+39
Fifty percent of this
1/2 of this
RESULT

Puzzle 60

36
Divide by three
1/3 of this
÷2
Multiply by five
+12
1/2 of this
RESULT

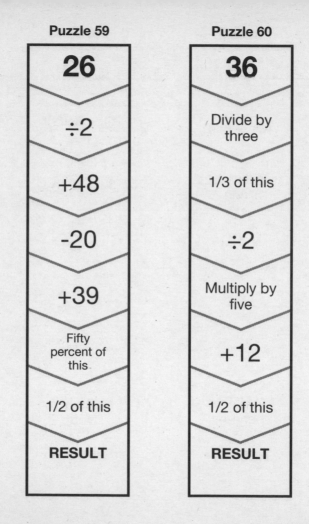

Puzzle 61

49

Add eleven

Divide by five

+44

-48

25% of this

Add fifty percent

RESULT

Puzzle 62

45

1/3 of this

Add seventeen

-14

1/2 of this

-7

×2

RESULT

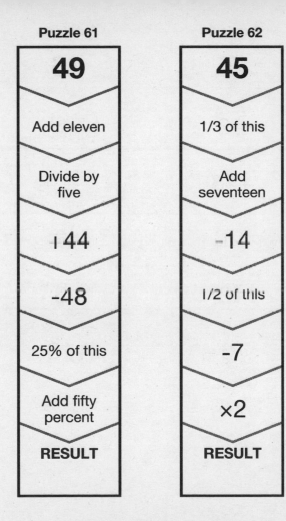

Puzzle 63

6

÷3

+21

-1

÷2

-6

×5

RESULT

Puzzle 64

26

1/2 of this

Add seventeen

÷6

×3

+20

Divide by five

RESULT

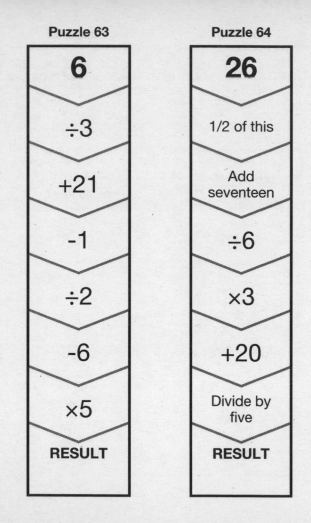

Puzzle 65

9
$\sqrt{\ }$
Multiply by ten
+6
-16
1/4 of this
+33
RESULT

Puzzle 66

6
-50%
Add forty-five
One third of this
÷8
+42
Divide by two
RESULT

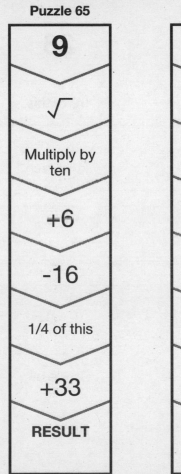

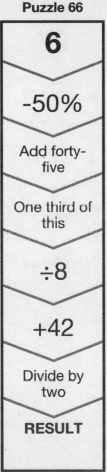

Puzzle 67

27
Divide by three
$\sqrt{\ }$
+35
50% of this
-8
+18
RESULT

Puzzle 68

38
1/2 of this
Multiply by three
+6
-7
50% of this
÷7
RESULT

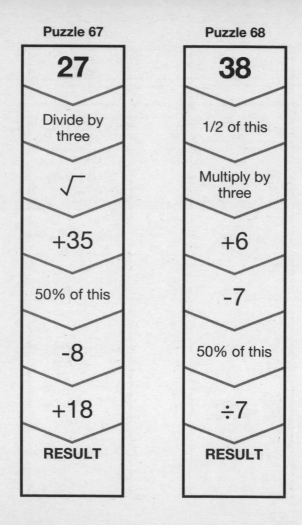

Puzzle 69

12

×2

1/4 of this

Fifty percent of this

+39

-10

+29

RESULT

Puzzle 70

27

÷9

Multiply by two

Divide by three

+35

Subtract one

Add twelve

RESULT

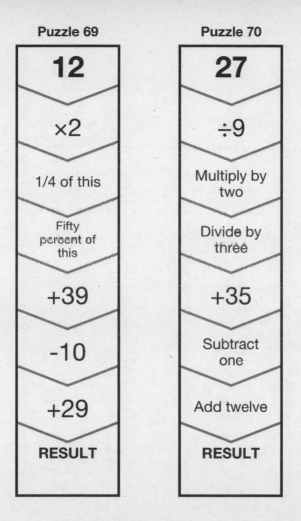

Puzzle 71

25
÷5
×12
One third of this
Add fourteen
Subtract one
÷3
RESULT

Puzzle 72

40
Subtract twenty-six
÷7
×4
75% of this
÷2
×11
RESULT

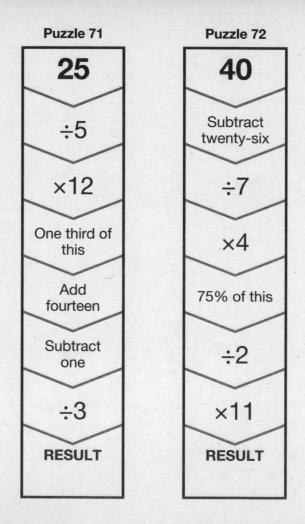

Puzzle 73

46
+5
÷3
×4
1/2 of this
+37
-8
RESULT

Puzzle 74

31
+1
1/4 of this
50% of this
$\sqrt{\ }$
×9
+50%
RESULT

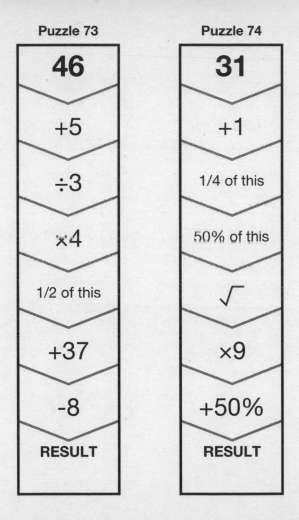

Puzzle 75

34

1/2 of this

-9

+12

50% of this

-50%

×5

RESULT

Puzzle 76

5

×9

Add thirty-four

-3

1/2 of this

50% of this

+27

RESULT

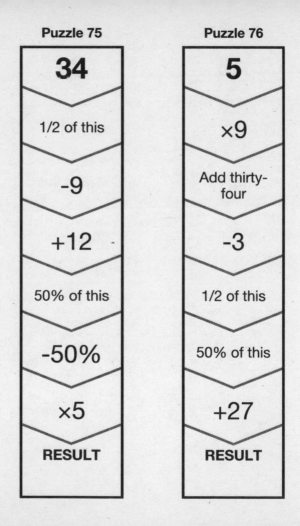

Puzzle 77

9

×2

+18

-30

×11

Fifty percent of this

1/3 of this

RESULT

Puzzle 78

27

+11

1/2 of this

+38

-7

Divide by two

×3

RESULT

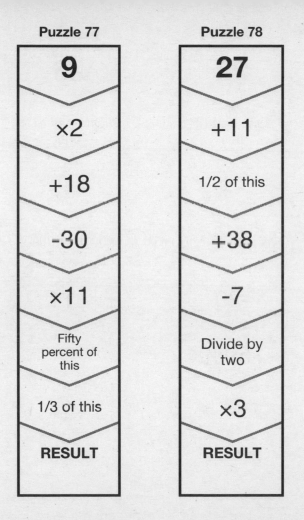

Puzzle 79

22
One half of this
-4
×11
-29
+50%
50% of this
RESULT

Puzzle 80

26
One half of this
+49
50% of this
+11
÷7
×9
RESULT

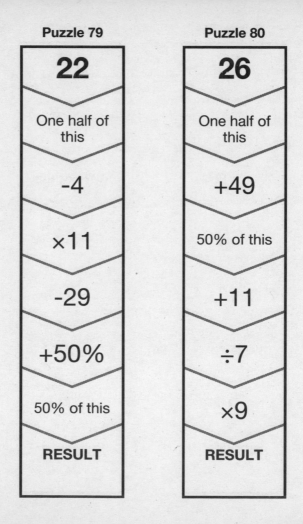

Puzzle 81

16
50% of this
Multiply by nine
Divide by two
-7
+2
-22
RESULT

Puzzle 82

15
Divide by three
60% of this
+6
×10
÷5
+70
RESULT

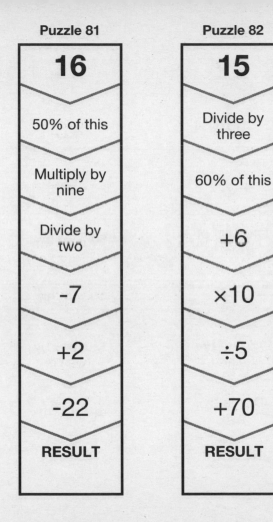

Puzzle 83

22
-50%
+63
One half of this
-18
Add thirty-nine
Divide by two
RESULT

Puzzle 84

31
+6
-25
50% of this
2/3 of this
Multiply by twelve
+50%
RESULT

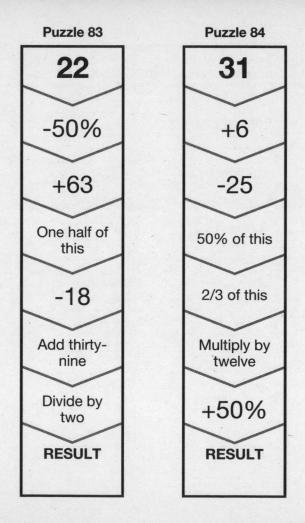

Puzzle 85

37
+29
1/2 of this
÷3
×2
Add seventeen
-18
RESULT

Puzzle 86

14
÷2
×12
1/2 of this
+12
÷3
One third of this
RESULT

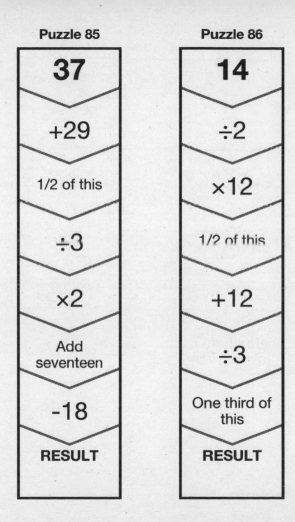

Puzzle 87

19
×2
+50%
Subtract twenty-six
+35
Subtract thirty-seven
Add fifty-six
RESULT

Puzzle 88

28
+7
-20
×6
Two thirds of this
Thirty percent of this
1/2 of this
RESULT

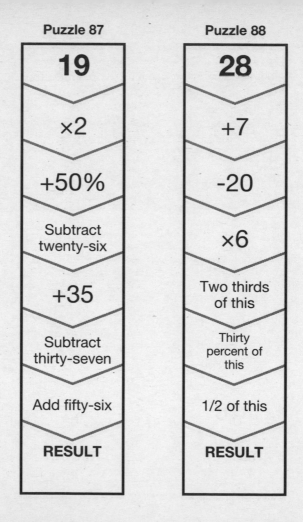

Puzzle 89

21
Two thirds of this
-50%
×5
80% of this
+40
-46
RESULT

Puzzle 90

23
Multiply by two
+50%
-36
÷11
+16
-1
RESULT

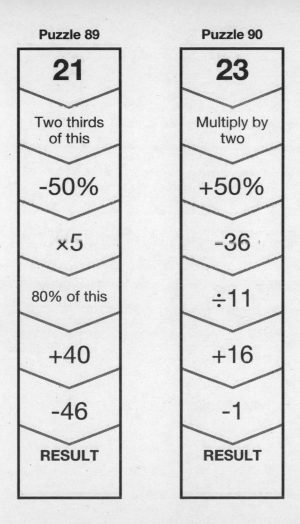

Puzzle 91

25

60% of this

Three fifths of this

÷3

+26

-5

50% of this

RESULT

Puzzle 92

36

÷3

Fifty percent of this

+9

Divide by five

×13

-34

RESULT

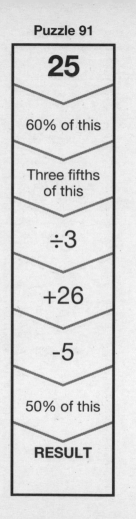

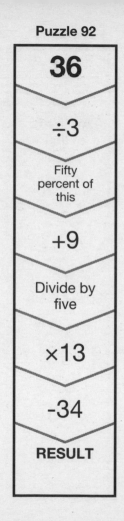

Puzzle 93

29
Add forty-nine
÷3
+71
-50
×2
-20
RESULT

Puzzle 94

47
×2
Subtract sixty-four
90% of this
Two thirds of this
50% of this
$\sqrt{}$
RESULT

Puzzle 95	Puzzle 96
43	**8**
Multiply by two	One half of this
-71	×12
+14	+30
-25	÷2
×8	-20
÷4	+45
RESULT	**RESULT**

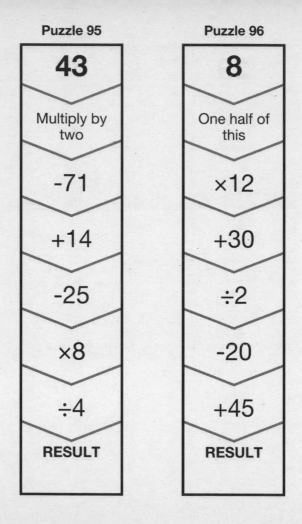

Puzzle 97

22
+23
Subtract thirty-three
1/2 of this
Add forty-four
-2
Fifty percent of this
RESULT

Puzzle 98

33
Add one
Fifty percent of this
+28
÷3
×4
-51
RESULT

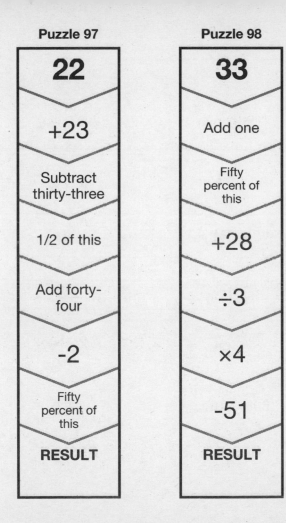

Puzzle 99

16

1/2 of this

Add twelve

70% of this

+11

÷5

80% of this

RESULT

Puzzle 100

20

1/2 of this

+44

2/3 of this

÷4

Add sixty-five

-14

RESULT

Puzzle 101

24
Fifty percent of this
×8
1/2 of this
+3
÷3
×2
RESULT

Puzzle 102

8
+70
÷13
Multiply by seven
+50%
Divide by nine
×5
RESULT

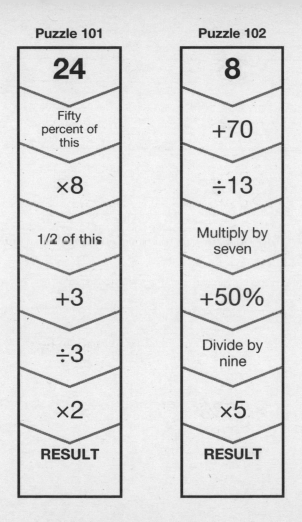

Puzzle 103

35

÷5

+13

÷4

80% of this

+24

+50%

RESULT

Puzzle 104

34

1/2 of this

×3

+32

-6

Divide by eleven

×12

RESULT

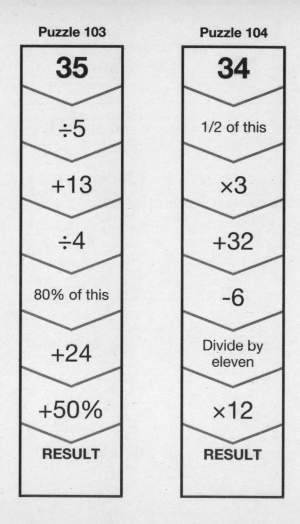

Puzzle 105

14

+31

3/5 of this

÷3

√

Multiply by eleven

Add twenty-seven

RESULT

Puzzle 106

6

×7

1/3 of this

÷2

+55

Subtract thirty-two

Divide by three

RESULT

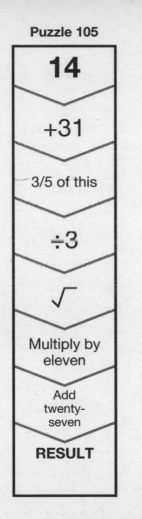

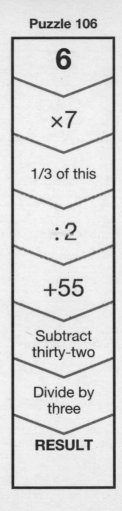

Puzzle 107

27

Subtract twenty-three

+45

÷7

+69

50% of this

1/2 of this

RESULT

Puzzle 108

45

Forty percent of this

÷2

Add eleven

×2

70% of this

Add sixteen

RESULT

Puzzle 109

38

1/2 of this

Add seventeen

50% of this

Add thirty-six

Subtract thirty

-50%

RESULT

Puzzle 110

43

-5

1/2 of this

-8

×2

+60

Fifty percent of this

RESULT

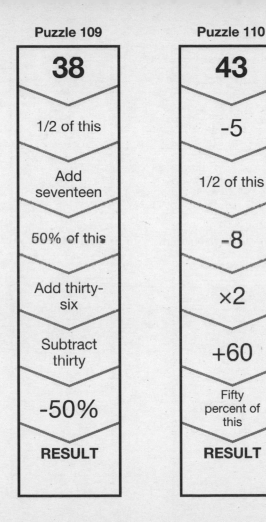

Puzzle 111

43

+48

÷13

Add eight

2/3 of this

×9

Divide by two

RESULT

Puzzle 112

24

-50%

×8

2/3 of this

$\sqrt{\ }$

+32

80% of this

RESULT

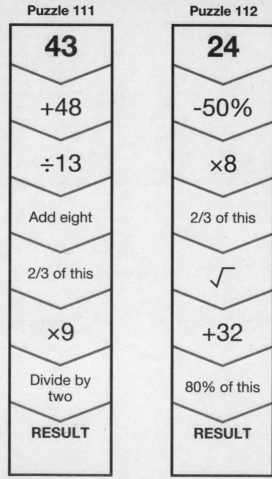

Puzzle 113

43
+55
-51
+30
-73
×8
Add fifty percent
RESULT

Puzzle 114

19
×5
-60%
+23
-25
+50%
1/2 of this
RESULT

Puzzle 115

20
1/2 of this
×8
40% of this
+27
-14
60% of this
RESULT

Puzzle 116

14
50% of this
+45
-3
+10
Subtract nine
3/5 of this
RESULT

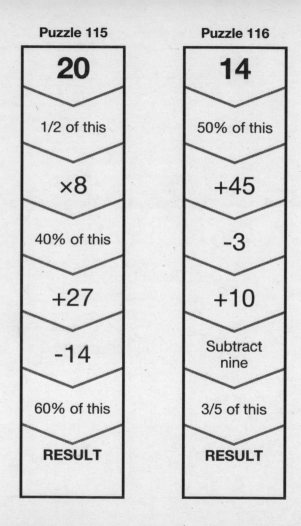

Puzzle 117

37

Multiply by two

-34

2/5 of this

$\sqrt{\ }$

×8

Subtract five

RESULT

Puzzle 118

37

-4

÷3

+34

÷9

×10

Add nineteen

RESULT

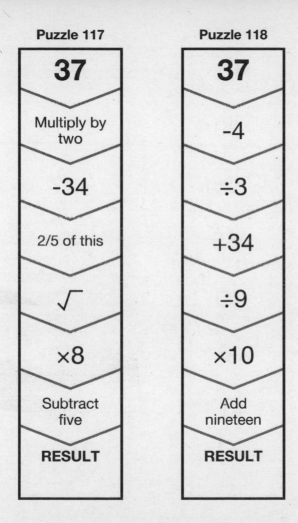

Puzzle 119	Puzzle 120
49	**44**
-39	1/2 of this
+70%	+37
+65	-2
-10	2/3 of this
÷6	÷2
+1	+58
RESULT	**RESULT**

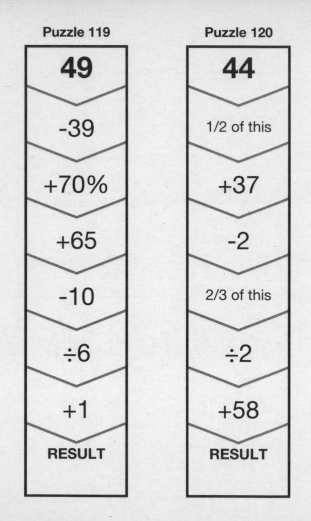

Puzzle 121

18

50% of this

×9

÷3

×2

+3

1/3 of this

RESULT

Puzzle 122

50

1/2 of this

÷5

×11

Subtract thirty-nine

50% of this

÷2

RESULT

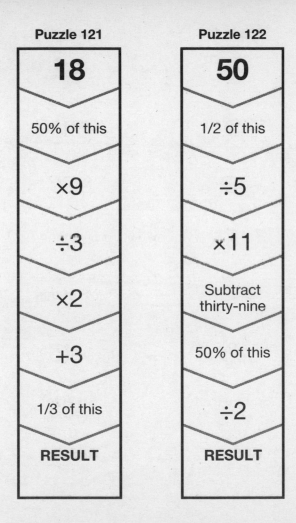

Puzzle 123

25
Divide by five
×3
Add eighty percent
1/3 of this
Multiply by four
Square root of this
RESULT

Puzzle 124

42
-50%
÷7
Add nine
1/2 of this
×15
Subtract thirty-one
RESULT

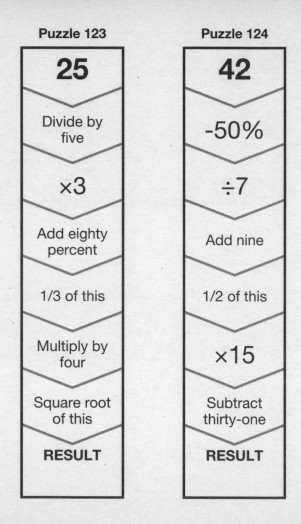

Puzzle 125

46
1/2 of this
-2
One third of this
Add seven
×6
Fifty percent of this
RESULT

Puzzle 126

34
÷2
+27
-9
÷5
×11
+15
RESULT

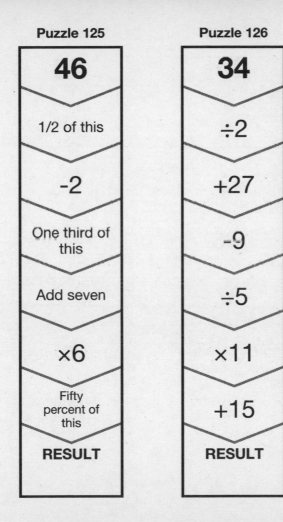

Puzzle 127

49
+8
÷3
+66
-37
50% of this
÷8
RESULT

Puzzle 128

10
1/2 of this
Add nineteen
÷6
+15
×5
-73
RESULT

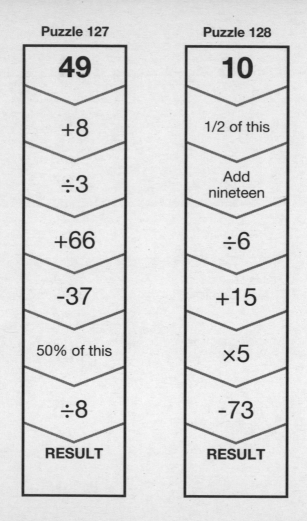

Puzzle 129

33

×2

÷11

×8

50% of this

+50%

+45

RESULT

Puzzle 130

46

Divide by two

×3

+8

÷11

+18

2/5 of this

RESULT

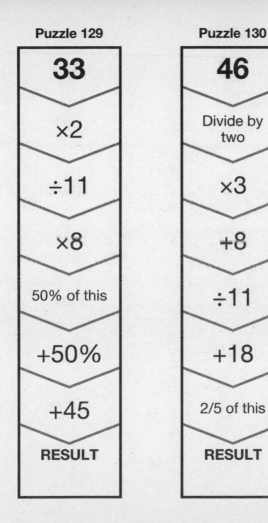

Puzzle 131

24
1/3 of this
+3
Multiply by five
-24
+43
÷2
RESULT

Puzzle 132

36
÷3
-50%
50% of this
+67
÷7
+52
RESULT

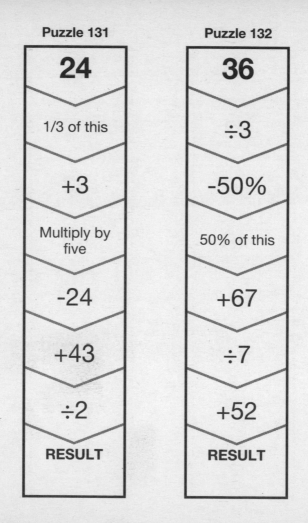

Puzzle 133

29
+32
Subtract twenty-one
÷5
×6
1/4 of this
Add forty-five
RESULT

Puzzle 134

26
+50%
Divide by three
-2
×6
-36
90% of this
RESULT

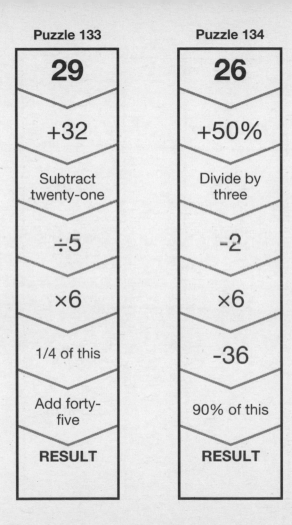

Puzzle 135

40
40% of this
+25
-34
×9
÷7
+14
RESULT

Puzzle 136

45
1/5 of this
+45
Subtract eight
50% of this
+72
2/5 of this
RESULT

Puzzle 137

34

÷2

+71

-37

+13

√

-50%

RESULT

Puzzle 138

22

+50%

+43

60% of this

÷2

-15

×6

RESULT

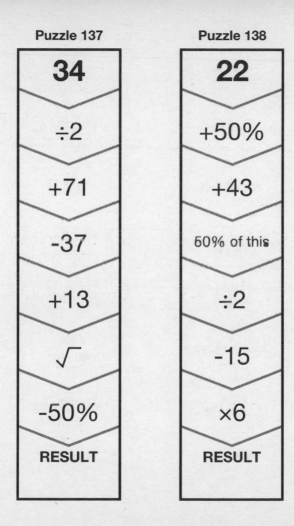

Puzzle 139

22

+13

3/5 of this

Add forty-three

-53

Multiply by nine

1/3 of this

RESULT

Puzzle 140

14

Divide by two

×10

÷7

1/2 of this

Eighty percent of this

Multiply by eight

RESULT

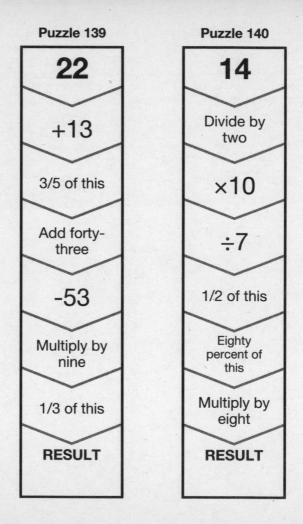

Puzzle 141

9
$\sqrt{}$
Add sixty-one
1/2 of this
Subtract fifty percent
+31
-28
RESULT

Puzzle 142

41
-1
+40%
+18
÷2
-32
Add fifteen
RESULT

Puzzle 143

38

-50%

+15

50% of this

+32

Divide by seven

×8

RESULT

Puzzle 144

10

+71

1/3 of this

-24

+40

-35

Divide by two

RESULT

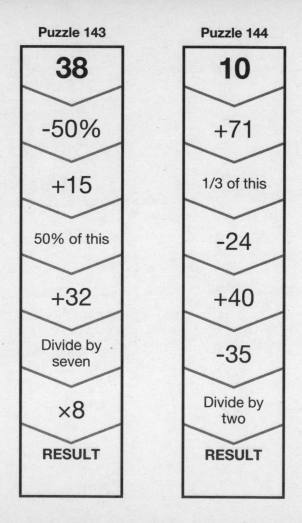

Puzzle 145

18
+5
×3
+14
-28
Divide by eleven
×2
RESULT

Puzzle 146

40
One fifth of this
-4
×13
1/2 of this
50% of this
+36
RESULT

Puzzle 147

35
Divide by seven
×3
Add twenty percent
50% of this
√
+7
RESULT

Puzzle 148

25
Eighty percent of this
+35
÷11
×7
-24
Add fifty-seven
RESULT

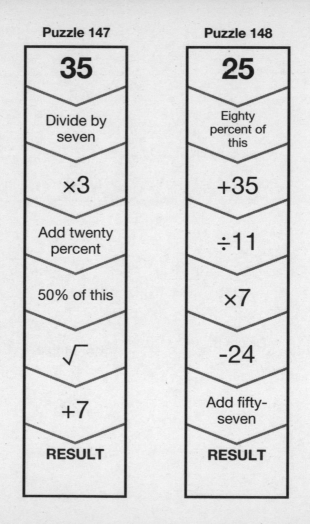

Puzzle 149

15
2/5 of this
50% of this
Multiply by four
+59
-9
One half of this
RESULT

Puzzle 150

24
÷8
+68
-60
×4
1/2 of this
+74
RESULT

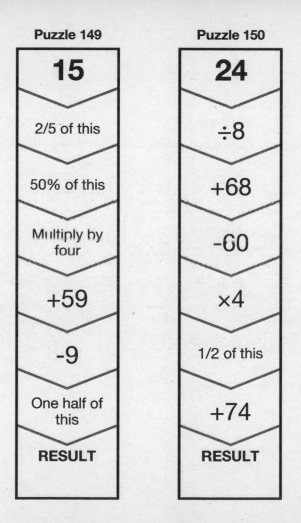

Puzzle 151

41
+33
-35
Add twenty-one
Thirty percent of this
One half of this
$\sqrt{\ }$
RESULT

Puzzle 152

26
1/2 of this
Add twenty-nine
Divide by two
One third of this
+28
-21
RESULT

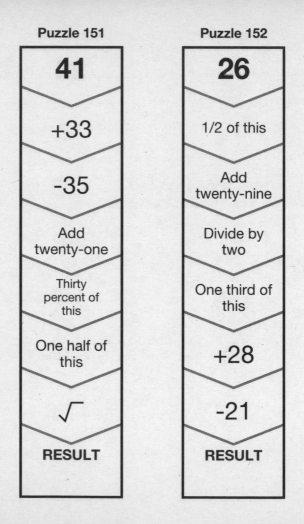

Puzzle 153

40
-25
-40%
$\sqrt{}$
×10
+51
-56
RESULT

Puzzle 154

50
3/5 of this
-26
+48
One half of this
×3
50% of this
RESULT

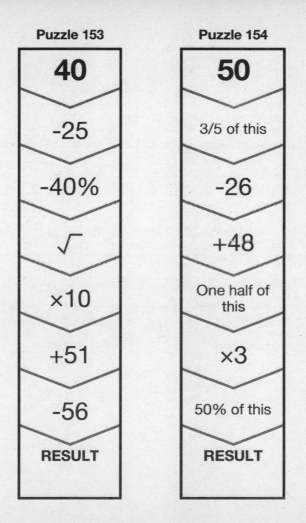

Puzzle 155

32

÷2

$\sqrt{}$

Add forty-five

Subtract twenty-two

+60

-68

RESULT

Puzzle 156

19

×4

One half of this

-34

+25

Subtract seven

Fifty percent of this

RESULT

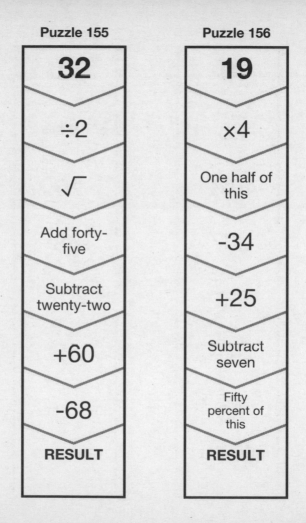

Puzzle 157

15
Eighty percent of this
+51
-37
1/2 of this
Add nineteen
Multiply by two
RESULT

Puzzle 158

25
÷5
×2
+41
One third of this
+11
-25
RESULT

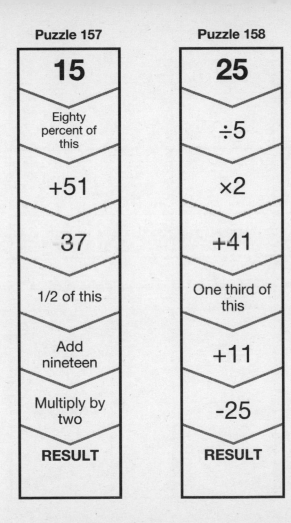

Puzzle 159

19

+13

÷2

+59

Three fifths of this

÷9

×5

RESULT

Puzzle 160

31

-9

1/2 of this

×3

÷11

+41

+50%

RESULT

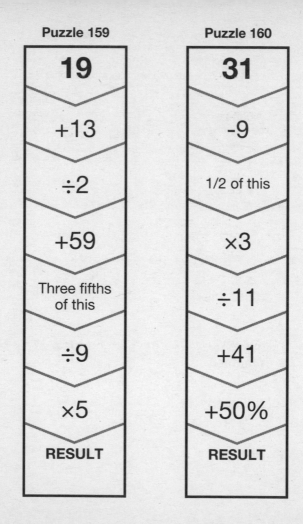

Puzzle 161

28
Subtract seventy-five percent
+69
-4
1/2 of this
Square root of this
×15
RESULT

Puzzle 162

34
Add forty-two
1/2 of this
Divide by two
-3
Add fifty-three
-47
RESULT

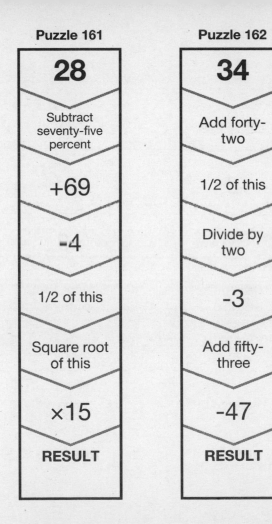

Puzzle 163

21

2/3 of this

+84

One half of this

÷7

Add fifty-two

Multiply by two

RESULT

Puzzle 164

81

Seven ninths of this

-51

×3

Square root of this

×5

÷2

RESULT

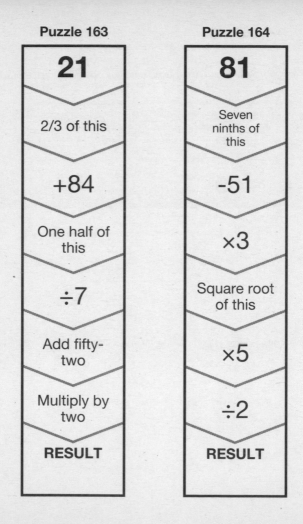

Puzzle 165

17
+55
1/2 of this
Divide by two
1/3 of this
+50%
+57
RESULT

Puzzle 166

14
1/2 of this
+8
1/3 of this
+42
×4
75% of this
RESULT

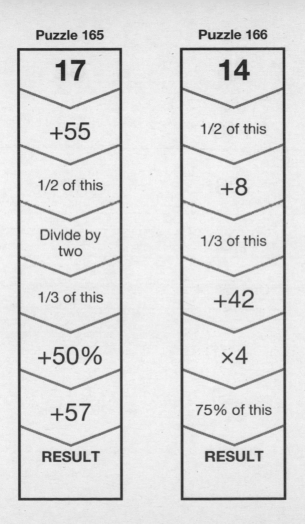

Puzzle 167

38
+10
-26
×6
One half of this
+39
÷7
RESULT

Puzzle 168

68
÷2
+50%
+85
-77
+60
1/7 of this
RESULT

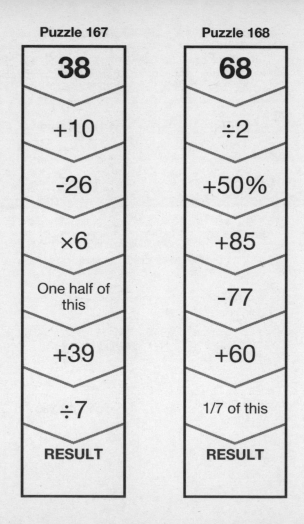

Puzzle 169

64
3/4 of this
Add ninety-seven
20% of this
×4
-35
Add sixty-nine
RESULT

Puzzle 170

24
+35
-25
÷2
+90
-55
1/2 of this
RESULT

Puzzle 171

21
1/3 of this
+55
1/2 of this
+25
÷14
×12
RESULT

Puzzle 172

87
1/3 of this
+23
÷2
Subtract fifteen
+59
40% of this
RESULT

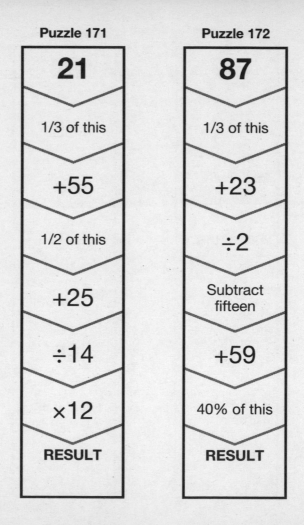

Puzzle 173

59
+33
÷4
+80
-27
×2
-26
RESULT

Puzzle 174

94
÷2
+4
1/3 of this
+66
-8
80% of this
RESULT

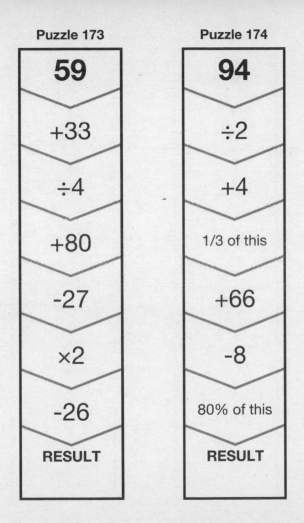

Puzzle 175

6
÷2
×12
1/9 of this
×16
Add ninety-nine
-2
RESULT

Puzzle 176

90
×2
-15
1/3 of this
-13
Fifty percent of this
+65
RESULT

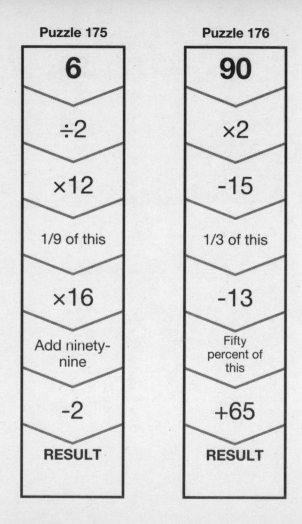

Puzzle 177

75

60% of this

+75

1/5 of this

÷4

Subtract fifty percent

×11

RESULT

Puzzle 178

6

×15

-45

20% of this

√

+65

-15

RESULT

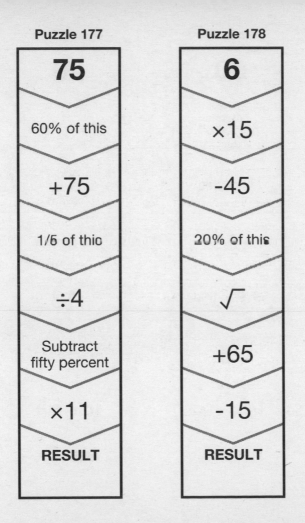

Puzzle 179

60
Fifteen percent of this
1/3 of this
+51
50% of this
×5
4/5 of this
RESULT

Puzzle 180

26
50% of this
Add seventy-seven
4/5 of this
-13
Add seven
1/2 of this
RESULT

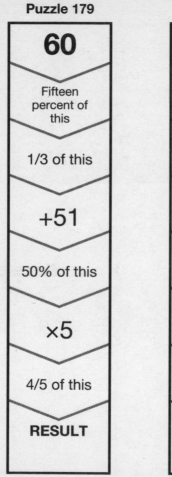

Puzzle 181

46
+4
Subtract thirty percent
Divide by five
×15
3/5 of this
-19
RESULT

Puzzle 182

12
Divide by three
Add sixty-six
-49
Divide by seven
×6
+1
RESULT

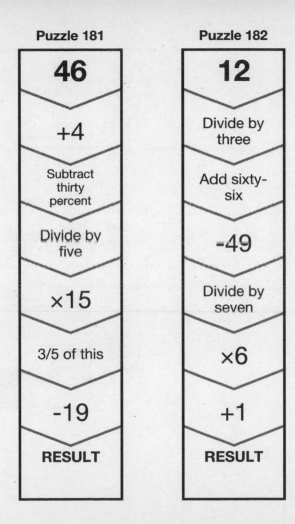

Puzzle 183

30
Add eighty-two
3/7 of this
-2
1/2 of this
-6
×9
RESULT

Puzzle 184

50
÷2
Four fifths of this
×6
+67
Divide by eleven
+71
RESULT

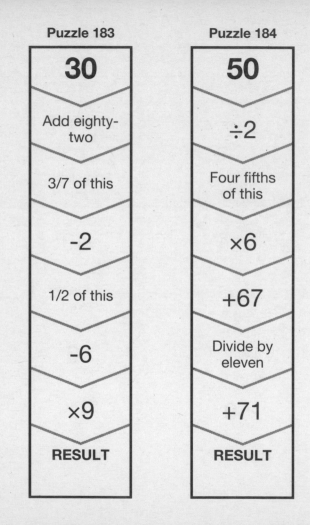

Puzzle 185

91

+47

Subtract twenty

1/2 of this

+2

×3

-60

RESULT

Puzzle 186

50

Subtract thirty

-10%

×2

$\sqrt{\ }$

÷2

Multiply by five

RESULT

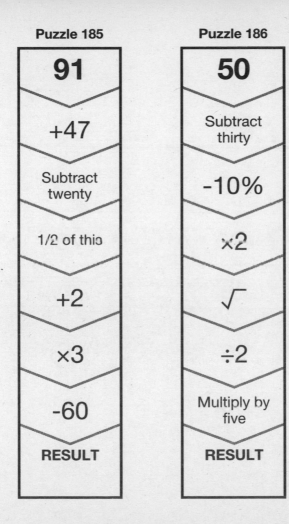

Puzzle 187

38
Subtract twenty-three
×5
-29
÷2
×6
+56
RESULT

Puzzle 188

58
1/2 of this
-23
×2
1/4 of this
×16
Divide by two
RESULT

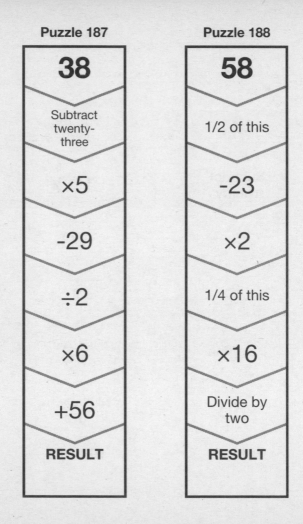

Puzzle 189

93
÷3
Subtract twenty
Add twenty-five
$\sqrt{\ }$
1/2 of this
×16
RESULT

Puzzle 190

31
-4
1/3 of this
÷3
+14
Subtract twelve
×13
RESULT

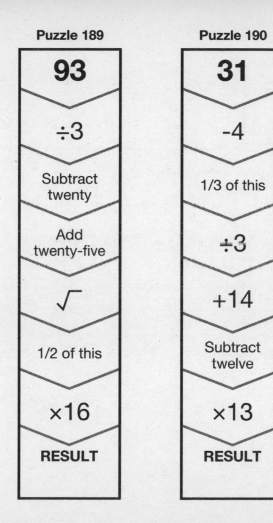

Puzzle 191

85
40% of this
×2
Add six
1/2 of this
Add eighty
7/9 of this
RESULT

Puzzle 192

39
+34
-11
+50%
Add ninety-one
Divide by four
1/2 of this
RESULT

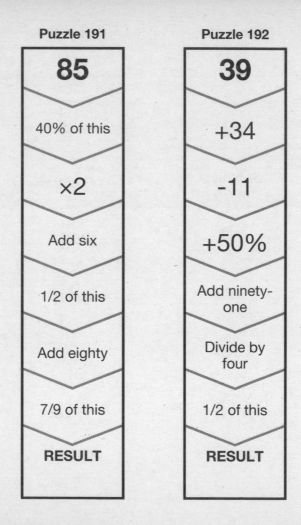

Puzzle 193

66
+39
One third of this
Forty percent of this
×8
÷7
+67
RESULT

Puzzle 194

75
40% of this
2/3 of this
÷4
+66
-38
+6
RESULT

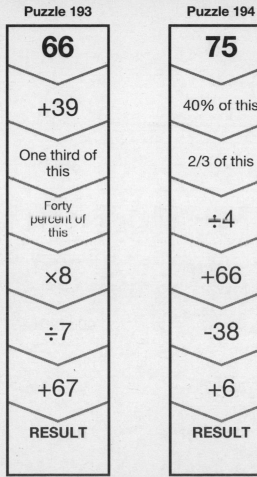

Puzzle 195

74
One half of this
×4
-94
One third of this
×7
+57
RESULT

Puzzle 196

88
-77
×15
÷11
Subtract forty percent
Add ninety-nine
Fifty percent of this
RESULT

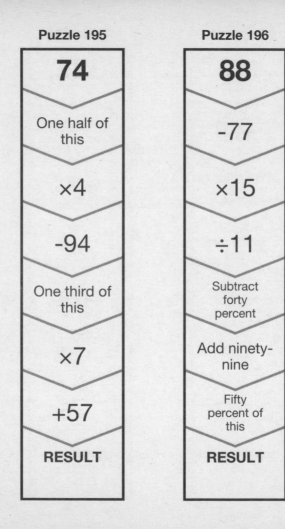

Puzzle 197

30

×6

Subtract thirty percent

50% of this

-46

×8

Subtract forty-one

RESULT

Puzzle 198

90

One third of this

-90%

Multiply by four

Seventy five percent of this

×16

+1

RESULT

Puzzle 199

89
×2
Subtract ninety-three
-60%
Multiply by four
-55
2/3 of this
RESULT

Puzzle 200

11
×11
+2
2/3 of this
÷2
+85
50% of this
RESULT

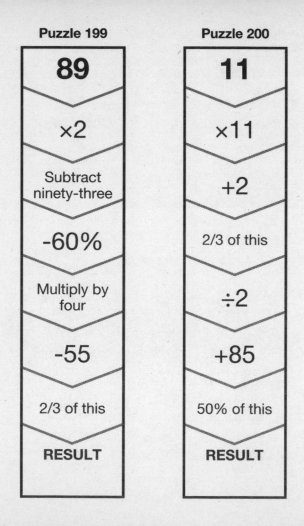

Puzzle 201

22
1/2 of this
+41
÷2
Multiply by three
+58
5/8 of this
RESULT

Puzzle 202

80
Add forty-five
20% of this
Subtract twenty-two
×4
50% of this
+39
RESULT

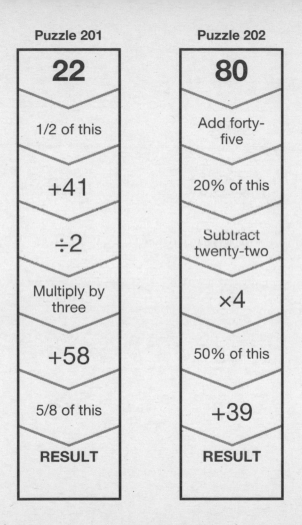

Puzzle 203

5
×13
+20%
-35
×4
-15
+42
RESULT

Puzzle 204

82
1/2 of this
+58
÷3
2/3 of this
50% of this
×13
RESULT

Puzzle 205

28

÷4

+44

÷3

-3

+91

1/3 of this

RESULT

Puzzle 206

84

1/6 of this

×10

+43

÷3

-2

Add eighty-two

RESULT

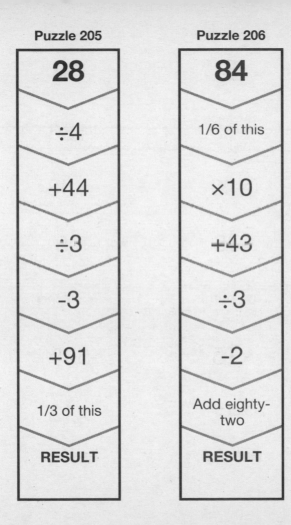

Puzzle 207

59
Subtract twenty-eight
+30
-46
20% of this
×15
2/3 of this
RESULT

Puzzle 208

19
+86
Three sevenths of this
÷5
2/3 of this
-50%
×12
RESULT

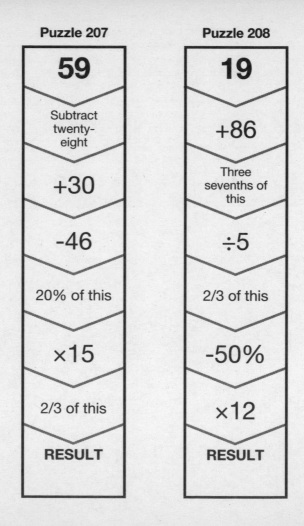

Puzzle 209

7
6/7 of this
×10
55% of this
÷3
+79
80% of this
RESULT

Puzzle 210

47
Subtract one
Divide by two
+11
1/2 of this
×9
+22
RESULT

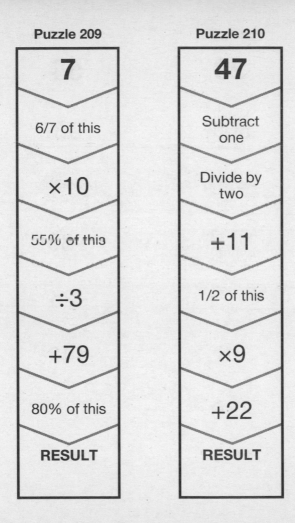

Puzzle 211

28
-50%
Add ten
Add fifty percent
One half of this
÷6
+59
RESULT

Puzzle 212

54
÷3
1/3 of this
+83
-26
+42
Divide by five
RESULT

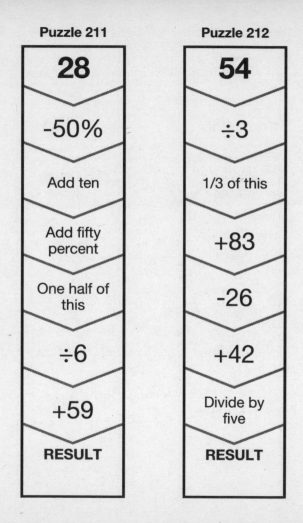

Puzzle 213

66
÷11
Subtract three
×7
Add ninety-three
2/3 of this
+28
RESULT

Puzzle 214

45
1/3 of this
Add thirty-five
Subtract twenty-three
+43
60% of this
1/2 of this
RESULT

Puzzle 215

84
Divide by three
×2
Six sevenths of this
+26
1/2 of this
+48
RESULT

Puzzle 216

59
-1
+50%
-20
+92
÷3
+40
RESULT

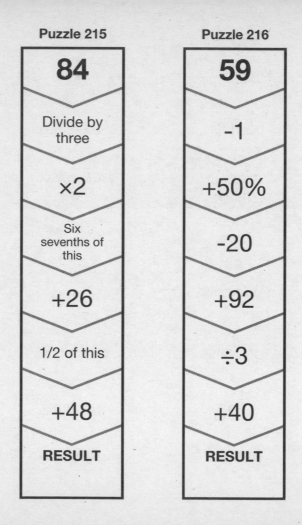

Puzzle 217

22
1/2 of this
×12
-50%
Add forty
Divide by two
-27
RESULT

Puzzle 218

42
50% of this
-16
×3
Add ten
Multiply by five
2/5 of this
RESULT

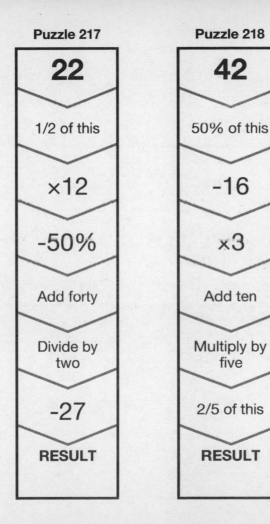

Puzzle 219

16
+95
1/3 of this
-32
+53
+50%
-67
RESULT

Puzzle 220

28
×2
÷7
+47
-17
+53
4/7 of this
RESULT

Puzzle 221

26
+50%
+4
-37
Add two
×10
-31
RESULT

Puzzle 222

61
-49
×4
50% of this
÷4
+84
Divide by two
RESULT

Puzzle 223

51

÷3

+36

-32

×3

÷7

+57

RESULT

Puzzle 224

38

Subtract twenty-one

×4

-24

+50

Divide by two

+28

RESULT

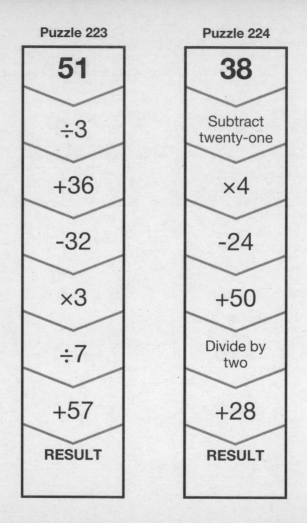

Puzzle 225

18
Subtract nine
+67
÷4
×9
-90
+83
RESULT

Puzzle 226

55
Divide by eleven
3/5 of this
+7
×5
+20%
-43
RESULT

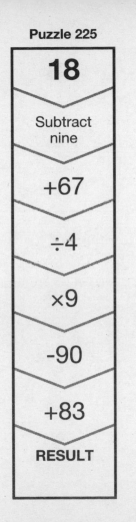

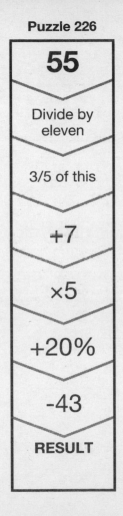

Puzzle 227

84
One half of this
÷3
+95
-20
+75
-54
RESULT

Puzzle 228

88
÷4
1/2 of this
Multiply by seventeen
-33
-50%
+24
RESULT

Puzzle 229

72
25% of this
+85
-38
80% of this
Divide by two
×7
RESULT

Puzzle 230

59
-33
-50%
Add seventy-five
Subtract seventy
÷3
Multiply by four
RESULT

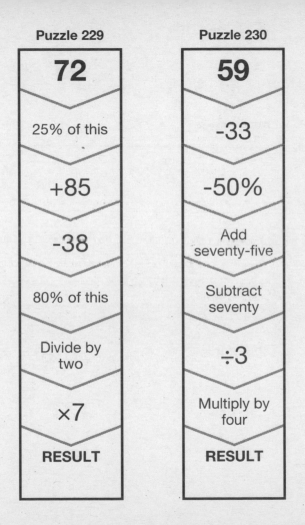

Puzzle 231

55

60% of this

÷11

+57

95% of this

2/3 of this

+18

RESULT

Puzzle 232

63

÷3

1/3 of this

×13

6/7 of this

-50%

+37

RESULT

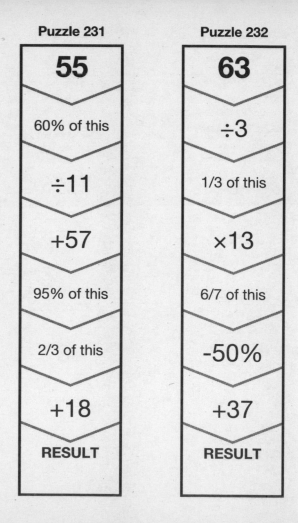

Puzzle 233

54
-39
+24
1/3 of this
×7
Add seventy-six
-89
RESULT

Puzzle 234

51
×3
Add thirty-seven
-22
75% of this
÷2
1/3 of this
RESULT

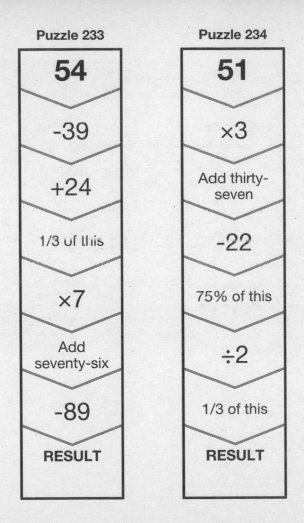

Puzzle 235

44

1/2 of this

50% of this

+50

Subtract thirty-five

÷2

+37

RESULT

Puzzle 236

69

1/3 of this

×7

+21

1/2 of this

÷7

+22

RESULT

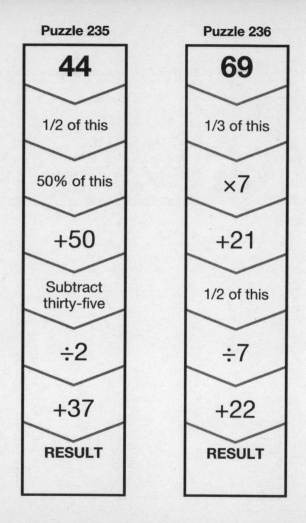

Puzzle 237

83

-72

+54

3/5 of this

÷3

×9

-53

RESULT

Puzzle 238

65

4/5 of this

-30

1/2 of this

Add eighty-nine

-44

75% of this

RESULT

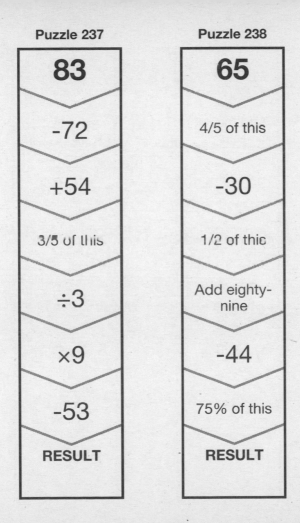

Puzzle 239

33
Add thirty-four
-45
50% of this
Multiply by twelve
1/2 of this
-31
RESULT

Puzzle 240

64
$\sqrt{\ }$
-50%
75% of this
+17
÷4
+25
RESULT

Puzzle 241

48
75% of this
+99
4/5 of this
+75
-53
70% of this
RESULT

Puzzle 242

88
One quarter of this
+15
-17
×5
1/2 of this
+72
RESULT

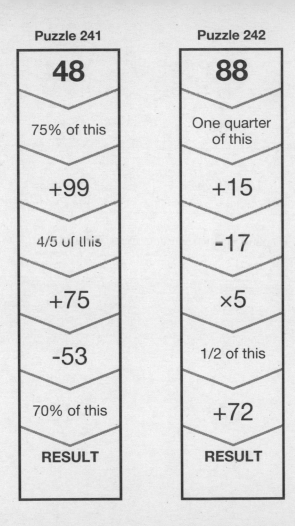

Puzzle 243

110
+68
1/2 of this
-58
×6
÷3
+7
RESULT

Puzzle 244

124
÷2
1/2 of this
-4
×6
÷9
×13
RESULT

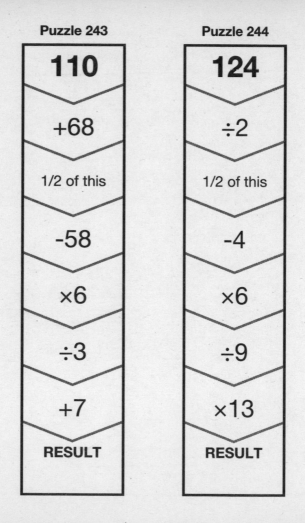

Puzzle 245

32
Fifty percent of this
×11
÷8
+96
+50%
-149
RESULT

Puzzle 246

40
+189
-130
1/3 of this
+162
-1
Subtract fifty percent
RESULT

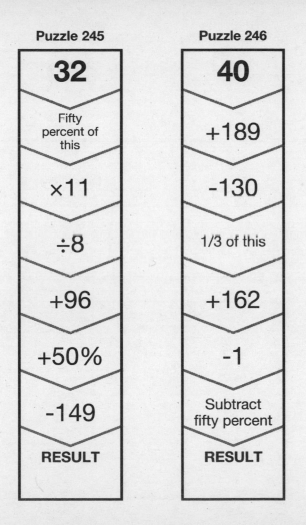

Puzzle 247

58
50% of this
+141
1/2 of this
+99
÷2
Subtract thirty
RESULT

Puzzle 248

129
+12
Divide by three
-3
×3
+94
-29
RESULT

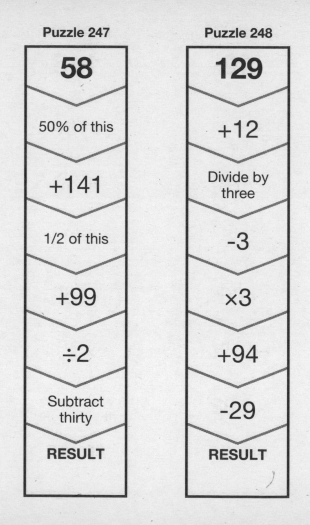

Puzzle 249

40
3/8 of this
+140
-25
1/2 of this
-40%
+198
RESULT

Puzzle 250

145
÷5
+19
Two thirds of this
Subtract fifty percent
×15
65% of this
RESULT

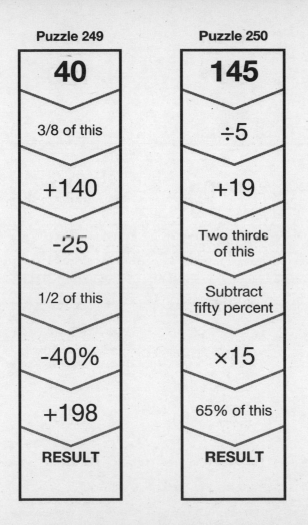

Puzzle 251

130
÷2
5/13 of this
+60%
+59
7/11 of this
-12
RESULT

Puzzle 252

30
5/6 of this
+42
-3
÷4
Multiply by ten
+75
RESULT

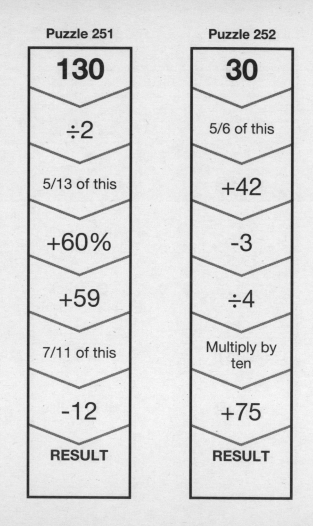

Puzzle 253

139
-8
+103
-50%
-37
+109
1/3 of this
RESULT

Puzzle 254

74
1/2 of this
+134
÷9
×2
+1
11/13 of this
RESULT

Puzzle 255

22
+50%
+76
Subtract forty
1/3 of this
Add one hundred and thirty-five
50% of this
RESULT

Puzzle 256

80
+124
-25%
2/9 of this
Subtract fifty percent
+176
Subtract one hundred and three
RESULT

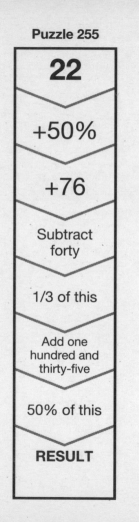

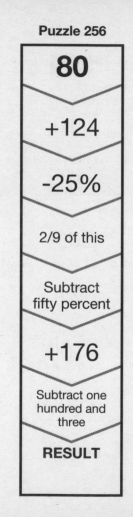

Puzzle 257

17
×10
-60%
1/2 of this
+84
Divide by two
+100
RESULT

Puzzle 258

52
1/2 of this
Multiply by five
+120
70% of this
÷7
+40%
RESULT

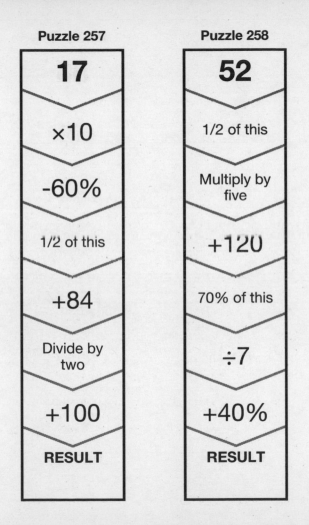

Puzzle 259

83

×2

+21

-67

70% of this

+54

-50%

RESULT

Puzzle 260

53

+38

×2

Divide by seven

+50%

+31

Subtract eight

RESULT

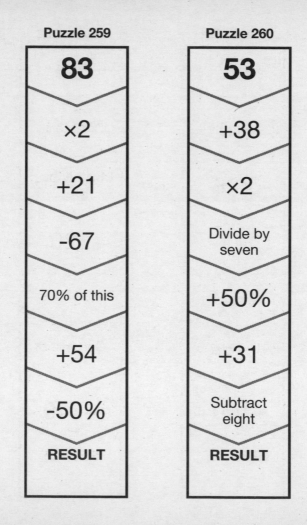

Puzzle 261

62

50% of this

+28

×2

+50%

-136

Add one hundred and eighty-seven

RESULT

Puzzle 262

93

2/3 of this

50% of this

Multiply by seven

-51

One half of this

+149

RESULT

Puzzle 263

141

Divide by three

+93

-74

Add thirty

1/2 of this

Multiply by three

RESULT

Puzzle 264

51

÷3

+128

-50

+132

Subtract fifty-five

÷2

RESULT

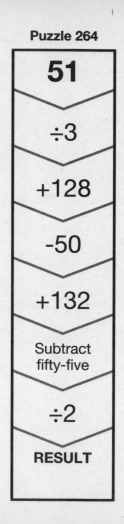

Puzzle 265

74
+44
-33
4/5 of this
÷4
×8
-94
RESULT

Puzzle 266

129
÷3
Add ninety-seven
+25%
2/5 of this
Add ten
×3
RESULT

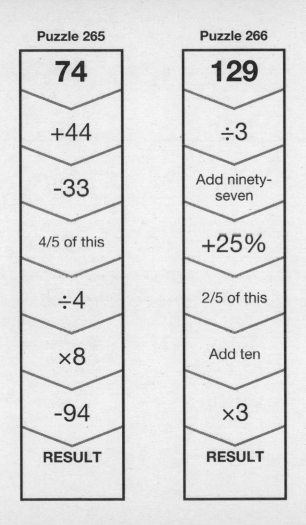

Puzzle 267

16
+25%
+26
Divide by two
×8
7/8 of this
Add one
RESULT

Puzzle 268

69
+38
×2
-120
1/2 of this
Add one hundred and fifty
-2
RESULT

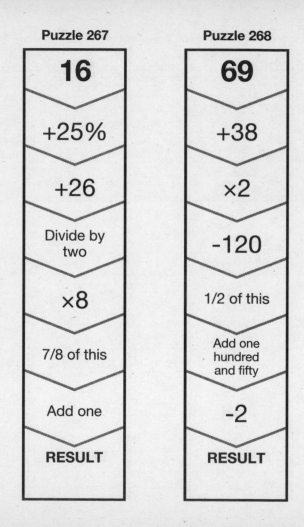

Puzzle 269

35

×6

Eighty percent of this

-56

Add forty-seven

-20

+58

RESULT

Puzzle 270

85

Add twenty

÷5

5/7 of this

Add one hundred and fifty-five

-36

1/2 of this

RESULT

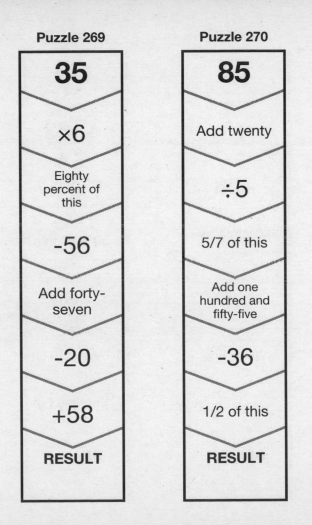

Puzzle 271

57
÷3
+28
Multiply by five
Forty percent of this
+153
2/13 of this
RESULT

Puzzle 272

83
+163
1/3 of this
50% of this
+69
1/2 of this
Multiply by four
RESULT

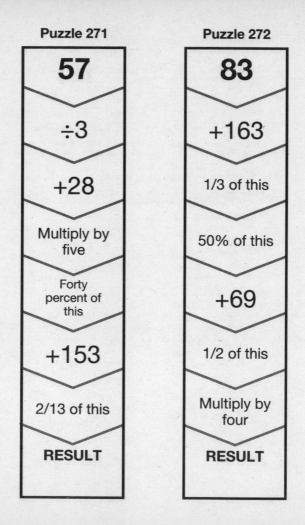

Puzzle 273

107
-88
+197
÷3
Seventy-five percent of this
+5
Multiply by three
RESULT

Puzzle 274

117
Subtract fifty-one
1/2 of this
+35
3/4 of this
+89
Subtract forty-seven
RESULT

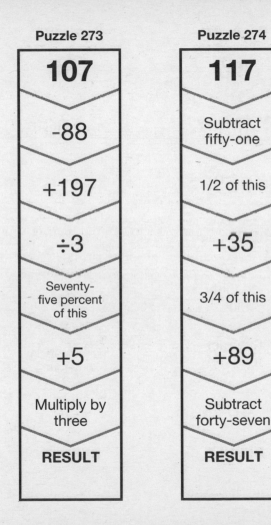

Puzzle 275

16
Multiply by eight
+25%
3/8 of this
Add thirty-nine
Subtract fourteen
+40%
RESULT

Puzzle 276

130
1/5 of this
+115
÷3
×5
Subtract one hundred and sixty-four
+34
RESULT

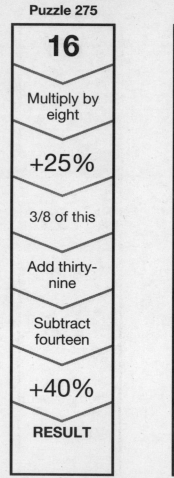

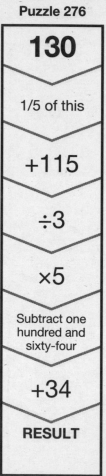

Puzzle 277

55
+73
Divide by four
Fifty percent of this
×9
Add ninety-nine
7/9 of this
RESULT

Puzzle 278

74
Divide by two
Add forty-five
1/2 of this
-18
×5
-48
RESULT

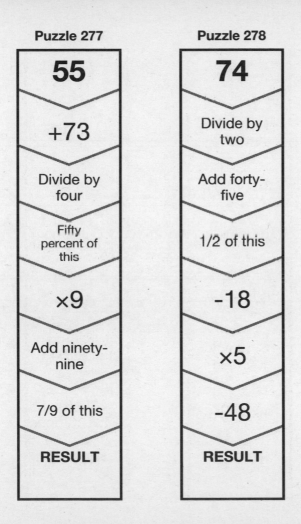

Puzzle 279

16
Multiply by twelve
1/2 of this
25% of this
3/4 of this
+132
÷5
RESULT

Puzzle 280

95
+96
-5
1/2 of this
÷3
+28
Subtract twenty-four
RESULT

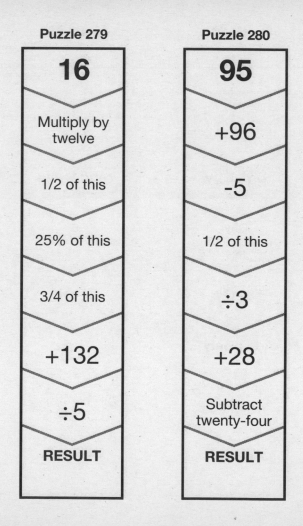

Puzzle 281

69

2/3 of this

Add one hundred and thirteen

Divide by three

+58

1/3 of this

-2

RESULT

Puzzle 282

26

Multiply by three

50% of this

+94

-98

+195

Three fifths of this

RESULT

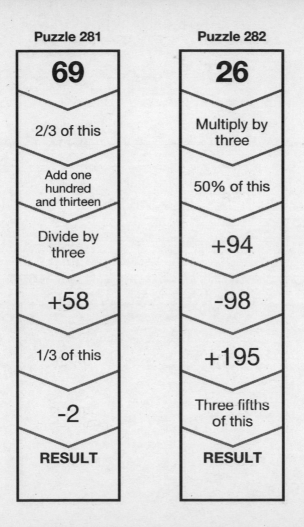

Puzzle 283

129
2/3 of this
÷2
+19
Subtract twenty-four
+68
1/2 of this
RESULT

Puzzle 284

90
Four ninths of this
-9
+69
75% of this
÷5
×12
RESULT

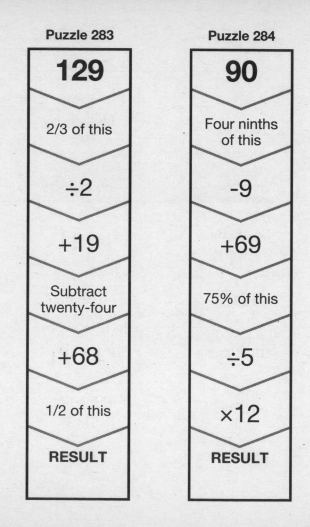

Puzzle 285

52
-50%
Multiply by five
Subtract fifteen
2/5 of this
+16
One half of this
RESULT

Puzzle 286

95
Three fifths of this
+189
2/3 of this
÷4
+138
-71
RESULT

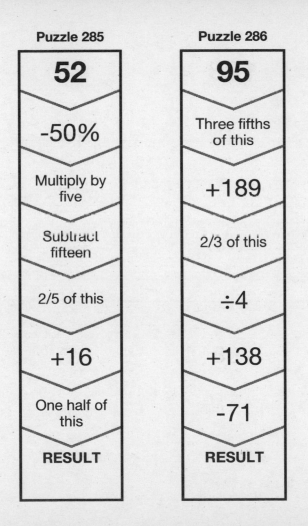

Puzzle 287

125

Add twenty percent

1/6 of this

+109

÷2

+57

Twenty-five percent of this

RESULT

Puzzle 288

100

÷4

+193

÷2

-88

×6

5/6 of this

RESULT

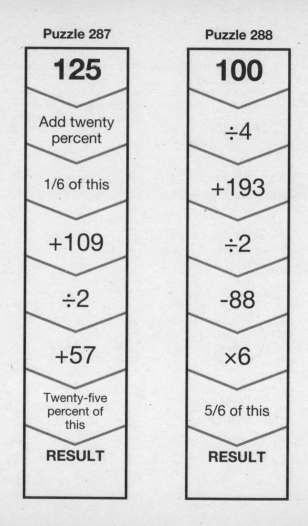

Puzzle 289

110

70% of this

+107

1/2 of this

÷4

+195

Subtract fifty percent

RESULT

Puzzle 290

140

1/2 of this

+98

-113

80% of this

+94

1/3 of this

RESULT

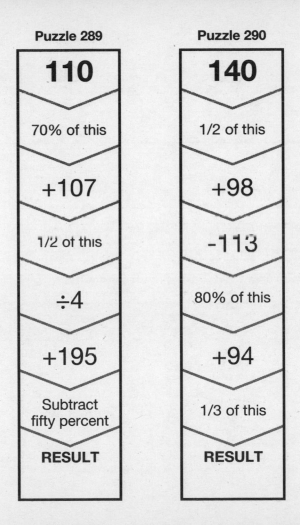

Puzzle 291

70
1/2 of this
+149
Twenty-five percent of this
+64
-80
+80%
RESULT

Puzzle 292

40
÷2
95% of this
+18
-5
1/2 of this
×14
RESULT

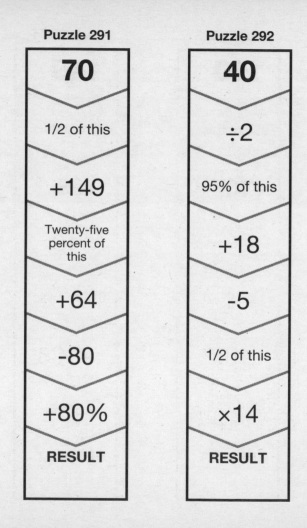

Puzzle 293

30
3/5 of this
Multiply by four
Seventy-five percent of this
+194
-192
Fifty percent of this
RESULT

Puzzle 294

49
3/7 of this
×6
÷3
2/3 of this
+128
-29
RESULT

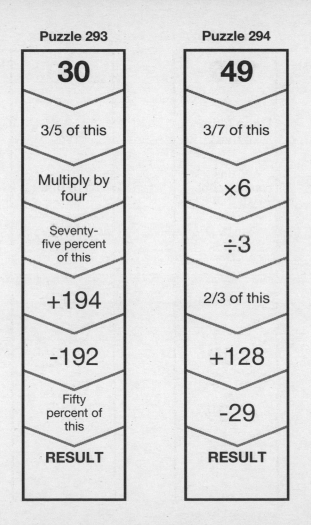

Puzzle 295

125
Three fifths of this
Add one hundred and sixty-five
÷8
50% of this
+102
2/3 of this
RESULT

Puzzle 296

41
×5
2/5 of this
+159
-102
+18
-6
RESULT

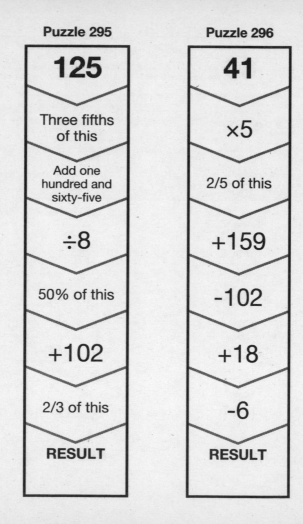

Puzzle 297

49
×3
3/7 of this
+50
-15
6/7 of this
+41
RESULT

Puzzle 298

112
+138
Subtract one hundred and seventy-four
+8
-50%
6/7 of this
75% of this
RESULT

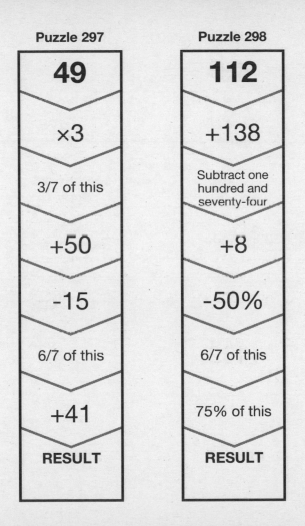

Puzzle 299

48

Divide by three

Add one hundred and fifty-three

4/13 of this

÷2

+145

1/3 of this

RESULT

Puzzle 300

15

+146

Subtract forty

Six elevenths of this

+106

1/2 of this

÷2

RESULT

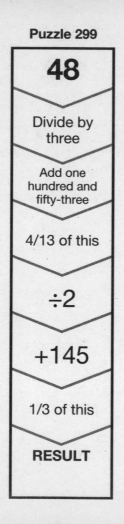

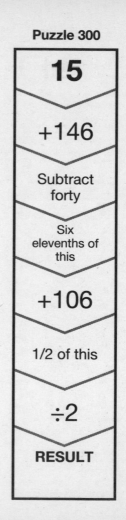

Puzzle 301

121

9/11 of this

÷3

+191

2/7 of this

+40

÷2

RESULT

Puzzle 302

80

+102

-54

Add oighty six

1/2 of this

Subtract three

50% of this

RESULT

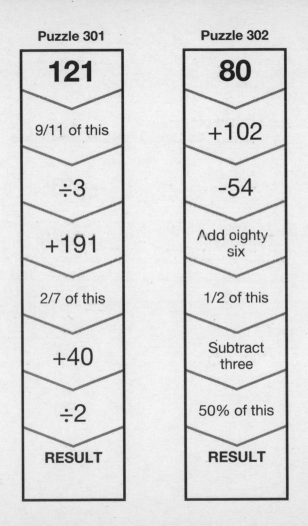

Puzzle 303

94
One half of this
+27
Fifty percent of this
×4
75% of this
Subtract seventy-seven
RESULT

Puzzle 304

53
-12
+43
1/2 of this
÷2
Subtract four
+70
RESULT

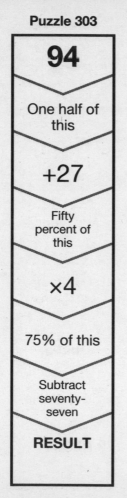

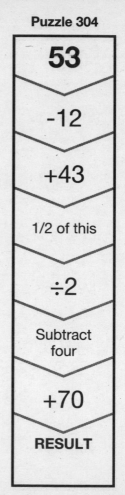

Puzzle 305

92
+75%
Divide by seven
+9
1/2 of this
+87
×2
RESULT

Puzzle 306

56
÷2
+38
50% of this
Subtract sixteen
+90
×2
RESULT

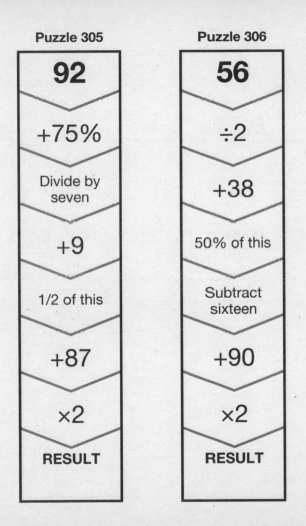

Puzzle 307

69
2/3 of this
×5
Divide by two
Add fifty-four
-29
95% of this
RESULT

Puzzle 308

97
-46
+105
÷3
One half of this
+115
1/3 of this
RESULT

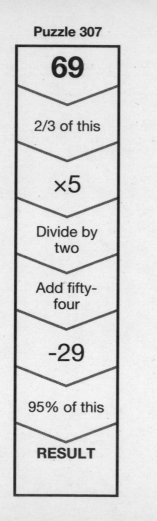

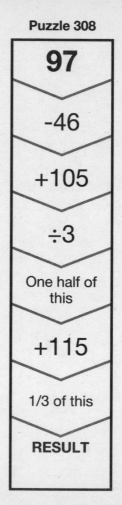

Puzzle 309

57

÷3

+15

50% of this

×8

-59

+157

RESULT

Puzzle 310

125

÷5

×6

-35

2/5 of this

+35

÷3

RESULT

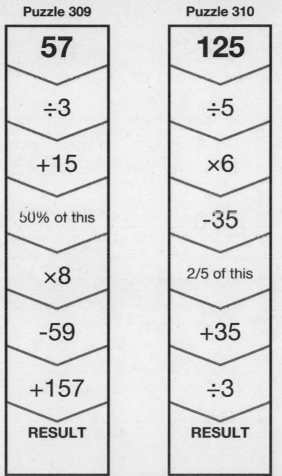

Puzzle 311

22
×10
÷11
+156
-89
+58
-87
RESULT

Puzzle 312

131
-81
×5
3/5 of this
Divide by six
+33
1/2 of this
RESULT

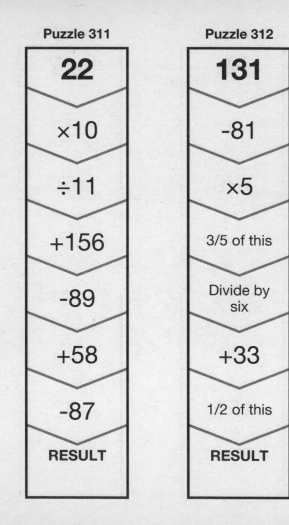

Puzzle 313

98
1/2 of this
+39
50% of this
Add one hundred and twenty-eight
÷2
-55
RESULT

Puzzle 314

117
×2
÷9
+195
9/13 of this
-55
+43
RESULT

Puzzle 315

80
-50
Add one hundred and ninety-nine
-175
50% of this
+138
Ten elevenths of this
RESULT

Puzzle 316

150
÷3
40% of this
+58
-26
50% of this
+108
RESULT

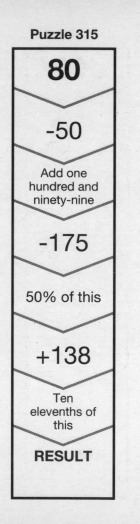

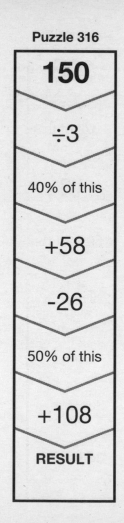

Puzzle 317

35
×5
3/5 of this
40% of this
2/3 of this
×4
Subtract four
RESULT

Puzzle 318

105
1/3 of this
+146
-103
-50%
+62
×2
RESULT

Puzzle 319

24
-25%
×4
1/2 of this
+165
-136
+60%
RESULT

Puzzle 320

123
×2
One third of this
Add forty-five
-98
×3
+71
RESULT

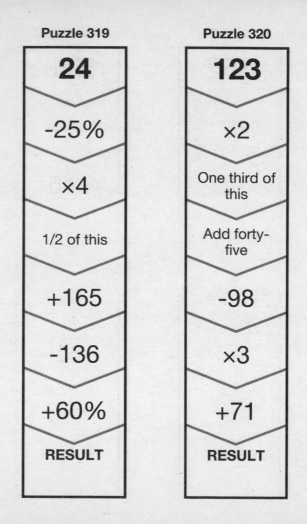

Puzzle 321

162
One half of this
+43
Divide by two
+50%
÷3
Add two hundred and fifty-six
RESULT

Puzzle 322

343
3/7 of this
+229
+50%
÷2
Add one hundred and twenty-seven
-122
RESULT

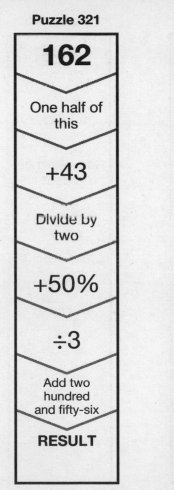

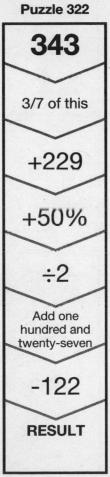

Puzzle 323

347

Add two hundred and thirty-two

2/3 of this

-98

1/2 of this

÷6

+279

RESULT

Puzzle 324

389

-329

Add twenty percent

Subtract forty

×18

$\sqrt{}$

Add two hundred and seventy-five

RESULT

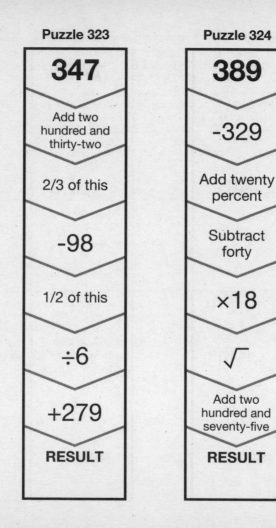

Puzzle 325

445
÷5
-15
+133
Divide by three
+194
-38
RESULT

Puzzle 326

148
Add eight
÷4
+5
One half of this
×5
-6
RESULT

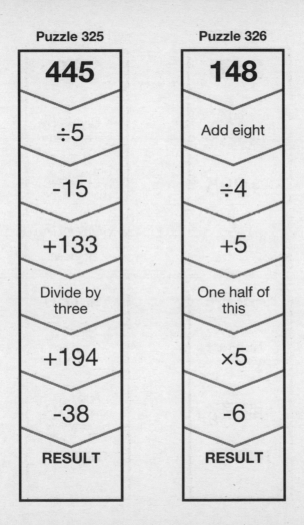

Puzzle 327

450
-129
+295
5/8 of this
-20%
Subtract twenty-five
+277
RESULT

Puzzle 328

396
Divide by eleven
×7
Subtract two hundred and thirteen
+120
÷3
+463
RESULT

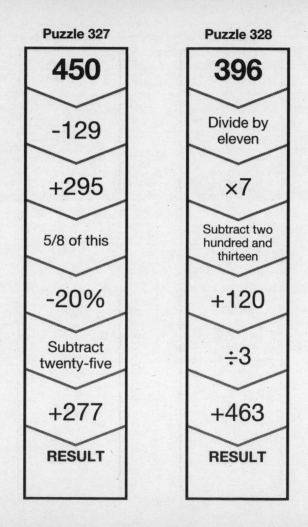

Puzzle 329

228
50% of this
+402
Divide by four
+94
×3
-74
RESULT

Puzzle 330

112
+362
÷2
2/3 of this
Add eighty-six
Divide by four
×10
RESULT

Puzzle 331

387
Divide by three
2/3 of this
-36
+314
1/7 of this
Add fifty
RESULT

Puzzle 332

86
-30
1/2 of this
Add three hundred and five
÷3
+210
2/3 of this
RESULT

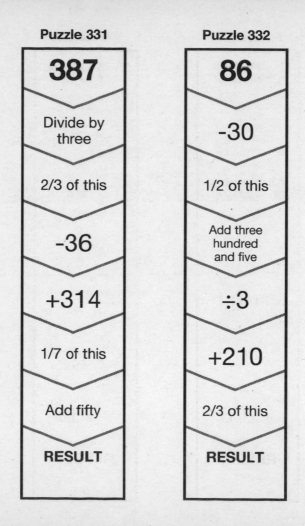

Puzzle 333

437

-333

Multiply by two

÷4

8/13 of this

×15

+208

RESULT

Puzzle 334

183

÷3

×12

Subtract sixty-three

1/3 of this

-32

+186

RESULT

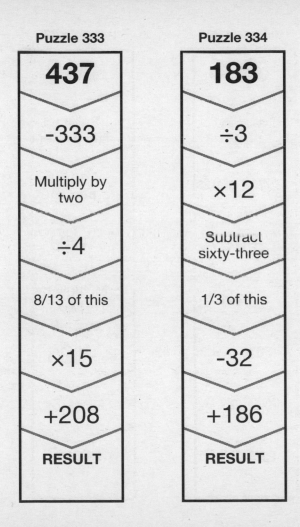

Puzzle 335

306

×2

7/12 of this

+387

÷3

50% of this

Multiply by four

RESULT

Puzzle 336

312

Subtract fifty percent

Add fifty percent

+49

Multiply by two

-193

+311

RESULT

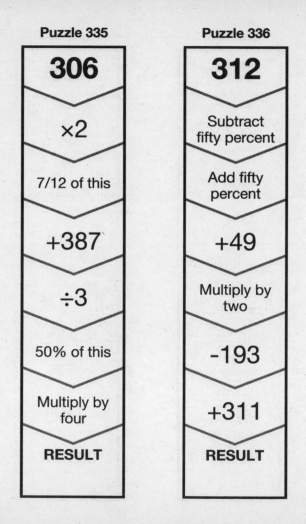

Puzzle 337

215
+73
1/3 of this
÷4
Multiply by seventeen
Subtract one hundred and seventy-five
+225
RESULT

Puzzle 338

488
÷8
+246
-255
50% of this
Add three hundred and sixty-six
1/2 of this
RESULT

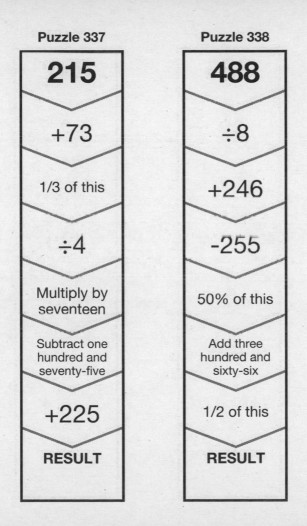

Puzzle 339

93

-32

+27

50% of this

÷2

+406

1/2 of this

RESULT

Puzzle 340

156

÷2

1/2 of this

+423

-430

Add four hundred and sixty-six

÷3

RESULT

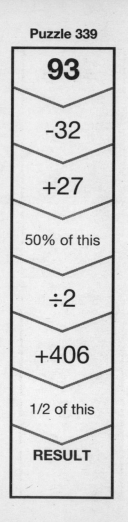

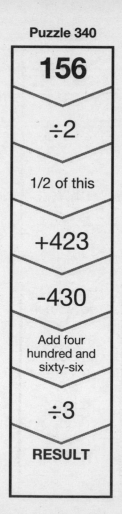

Puzzle 341

148

50% of this

1/2 of this

Add three hundred and twenty-nine

Two thirds of this

+160

-50%

RESULT

Puzzle 342

238

50% of this

+158

-119

+231

-183

÷2

RESULT

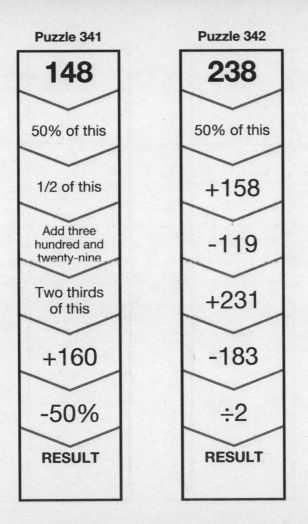

Puzzle 343

350
-46
Divide by four
+392
5/6 of this
-352
+211
RESULT

Puzzle 344

208
÷2
+398
50% of this
-221
70% of this
×12
RESULT

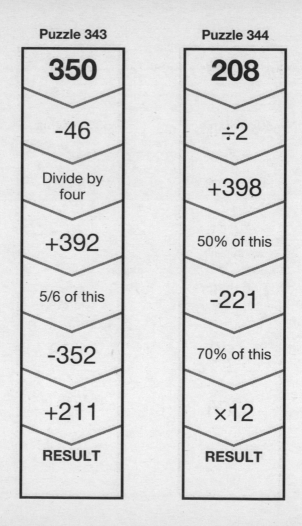

Puzzle 345

100
-2%
1/2 of this
+306
÷5
×10
-364
RESULT

Puzzle 346

498
1/3 of this
÷2
+162
Subtract fourteen
+68
12/13 of this
RESULT

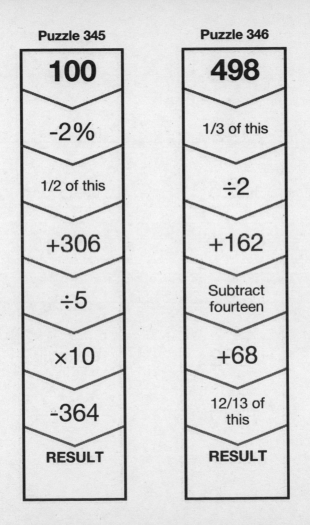

Puzzle 347

318
-142
+226
1/3 of this
÷2
+208
Multiply by two
RESULT

Puzzle 348

490
1/2 of this
÷5
+173
+50%
-220
×4
RESULT

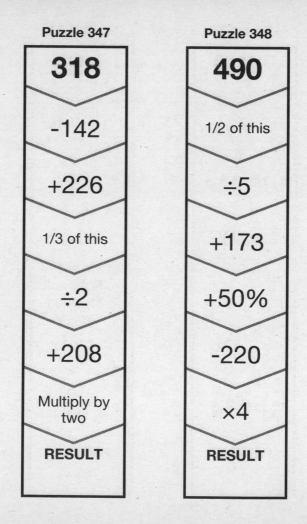

Puzzle 349

170
Sixty percent of this
÷2
+427
1/2 of this
-198
+186
RESULT

Puzzle 350

322
÷7
+179
2/9 of this
×14
Divide by five
+135
RESULT

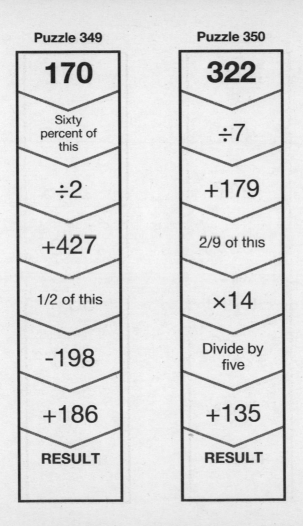

Puzzle 351

358
-323
×17
-60%
+440
5/6 of this
-361
RESULT

Puzzle 352

412
÷4
+245
÷12
×13
+237
1/2 of this
RESULT

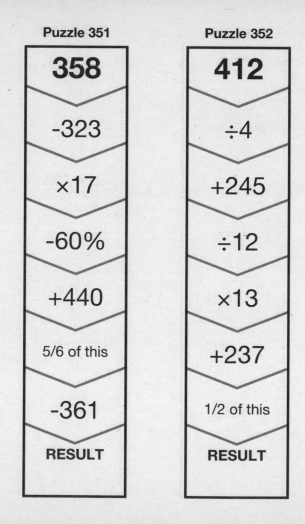

Puzzle 353

77

2/7 of this

×3

+50%

+365

5/8 of this

-35

RESULT

Puzzle 354

122

1/2 of this

+199

-118

Divide by two

×7

Four sevenths of this

RESULT

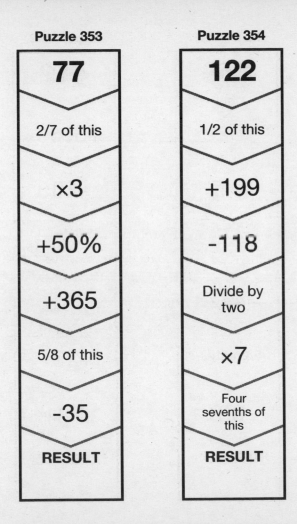

Puzzle 355

424

50% of this

÷4

-25

+447

4/5 of this

-120

RESULT

Puzzle 356

90

+80%

-10

Add four hundred and sixty-five

Subtract forty-five

Fifty percent of this

+231

RESULT

Puzzle 357

424
1/2 of this
÷4
×2
+3
Multiply by six
-64
RESULT

Puzzle 358

75
2/3 of this
-6
×6
÷2
-66
×5
RESULT

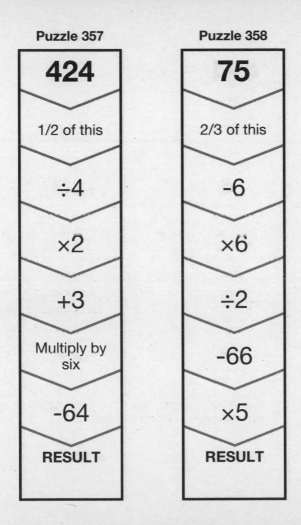

Puzzle 359

- **341**
- 6/11 of this
- ÷6
- ×18
- ÷9
- +203
- Subtract nineteen
- **RESULT**

Puzzle 360

- **382**
- -292
- +106
- -50%
- 1/2 of this
- ×9
- 1/3 of this
- **RESULT**

Puzzle 361

177

Divide by three

×8

÷4

+205

4/17 of this

Add forty-one

RESULT

Puzzle 362

337

Subtract one hundred and fifty-one

÷3

+379

√

Multiply by twenty-six

-434

RESULT

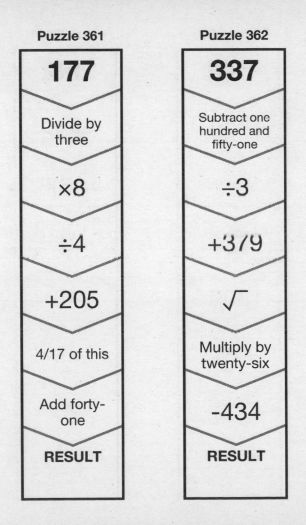

Puzzle 363

121
+220
5/11 of this
Divide by five
+309
-124
1/2 of this
RESULT

Puzzle 364

298
1/2 of this
+367
÷12
+48
2/7 of this
×23
RESULT

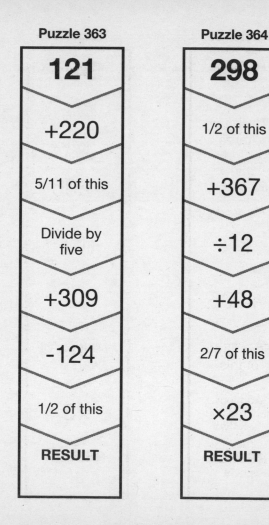

Puzzle 365

179
-84
2/5 of this
+462
÷4
Four fifths of this
Add four hundred and thirty-two
RESULT

Puzzle 366

345
÷5
+235
1/2 of this
÷2
+61
×3
RESULT

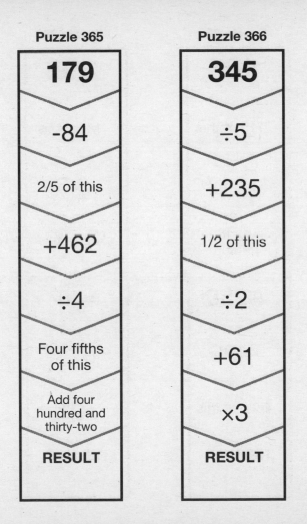

Puzzle 367

165
7/15 of this
+217
Add fifty percent
+250
-426
40% of this
RESULT

Puzzle 368

298
1/2 of this
×5
4/5 of this
Add seventy-three
Divide by three
-18
RESULT

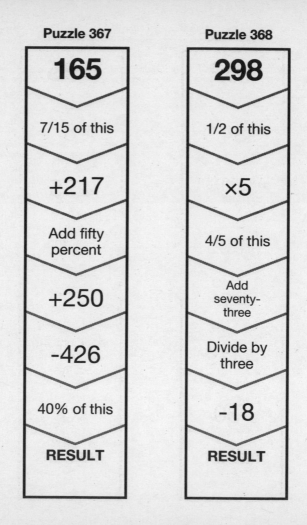

Puzzle 369

182
4/7 of this
Divide by four
+387
÷7
+419
-30
RESULT

Puzzle 370

495
2/5 of this
+55
Divide by eleven
×19
Eight nineteenths of this
×3
RESULT

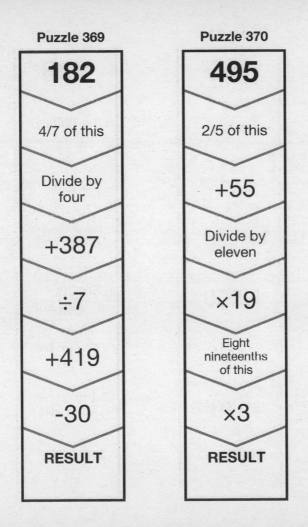

Puzzle 371

483
-148
80% of this
Divide by two
+222
÷4
+279
RESULT

Puzzle 372

166
One half of this
+77
Seventeen twentieths of this
Subtract fifty percent
+161
Subtract one hundred and nine
RESULT

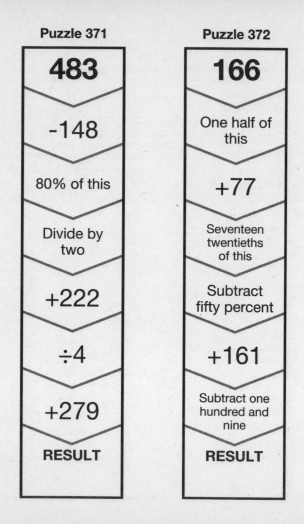

Puzzle 373

367

+198

80% of this

÷4

Add two hundred and eighteen

-127

2/3 of this

RESULT

Puzzle 374

319

3/11 of this

+93

40% of this

+423

÷15

×10

RESULT

Puzzle 375	Puzzle 376
303	**255**
+229	+316
-155	-373
÷13	2/3 of this
+413	-50%
÷2	÷3
13/17 of this	Add four hundred and eighty-nine
RESULT	**RESULT**

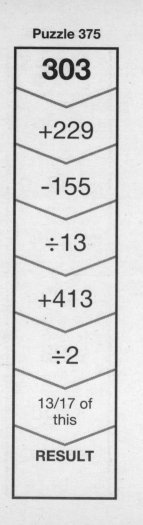

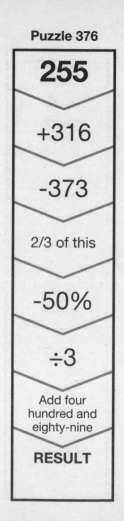

Puzzle 377

156
Divide by three
Multiply by ten
40% of this
÷2
×7
-118
RESULT

Puzzle 378

192
+179
÷7
×5
3/5 of this
-23
+438
RESULT

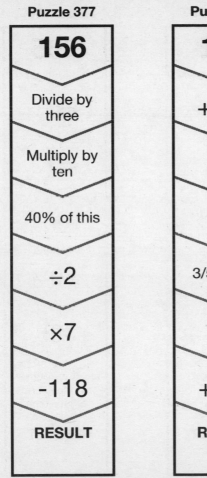

Puzzle 379

302
1/2 of this
+58
12/19 of this
50% of this
÷2
Add four hundred and two
RESULT

Puzzle 380

382
-87
÷5
+65
÷2
+495
-102
RESULT

Puzzle 381

150

+40

1/2 of this

-52

+415

-204

÷2

RESULT

Puzzle 382

127

+361

-72

÷4

1/2 of this

+21

Multiply by three

RESULT

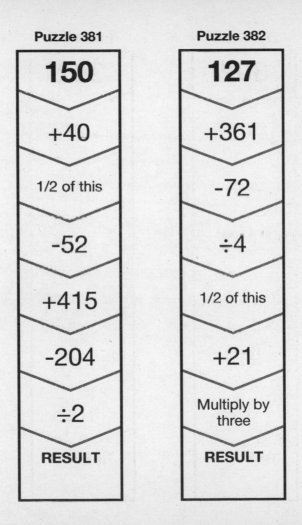

Puzzle 383

187
+39
1/2 of this
+153
-180
50% of this
×15
RESULT

Puzzle 384

259
-233
+153
-77
1/3 of this
+2
×13
RESULT

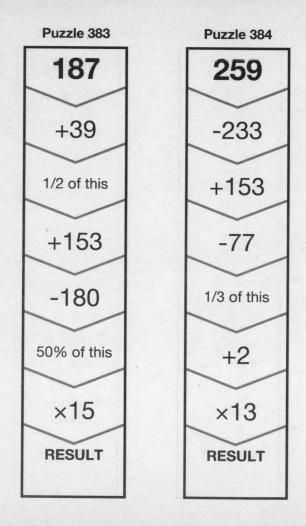

Puzzle 385

273
8/13 of this
Fifty percent of this
÷3
+411
Subtract ninety-nine
1/2 of this
RESULT

Puzzle 386

110
+379
÷3
-71
×7
Divide by twenty-three
Multiply by eleven
RESULT

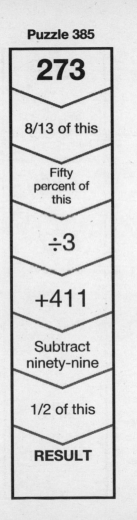

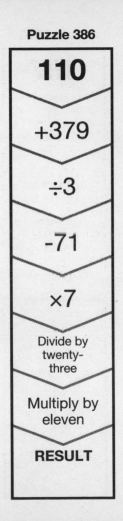

Puzzle 387

305

-13

1/2 of this

+36

50% of this

Three sevenths of this

Add seventy-seven

RESULT

Puzzle 388

233

Add four hundred and thirty-three

-376

÷10

×20

Subtract fourteen

÷2

RESULT

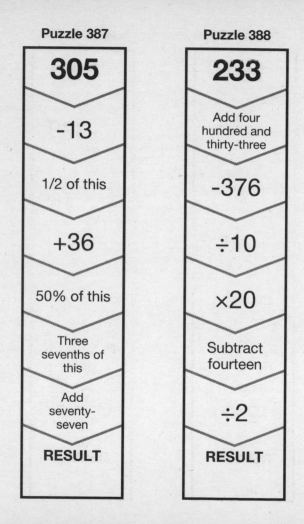

Puzzle 389

151
+453
÷2
+247
-4
Four fifths of this
-217
RESULT

Puzzle 390

62
+58
20% of this
×3
50% of this
Multiply by nine
1/3 of this
RESULT

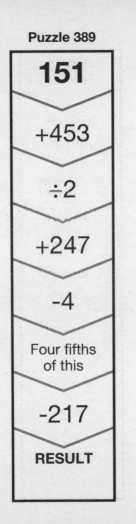

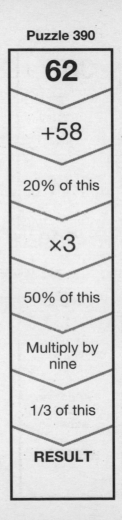

Puzzle 391

167

Subtract seventy-two

+144

-209

×25

One half of this

Add three hundred and sixty-two

RESULT

Puzzle 392

224

Add fifty-seven

-139

+439

3/7 of this

-18

2/3 of this

RESULT

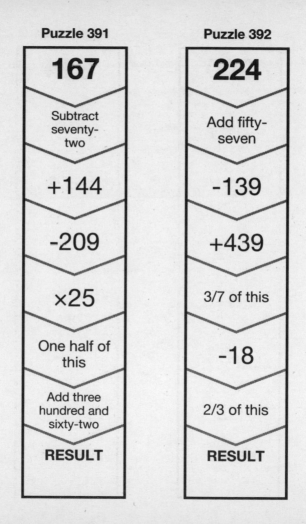

Puzzle 393

300
-163
+396
Nine thirteenths of this
+340
-231
÷2
RESULT

Puzzle 394

314
1/2 of this
+76
-164
×8
÷4
Add fifty percent
RESULT

Puzzle 395

296

1/2 of this

50% of this

+152

-141

3/5 of this

+80

RESULT

Puzzle 396

163

Add four hundred and seventy-nine

2/3 of this

÷2

-56

One half of this

+87

RESULT

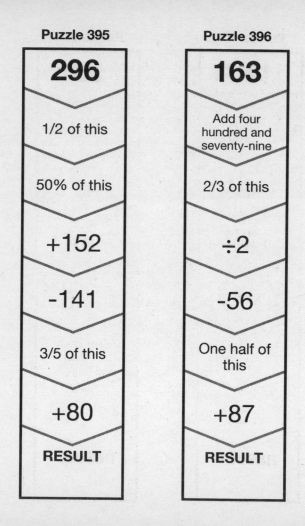

Puzzle 397

144
÷3
+239
5/7 of this
+223
One half of this
÷2
RESULT

Puzzle 398

53
×14
÷2
Add three hundred and two
-230
Add one hundred and fifteen
4/9 of this
RESULT

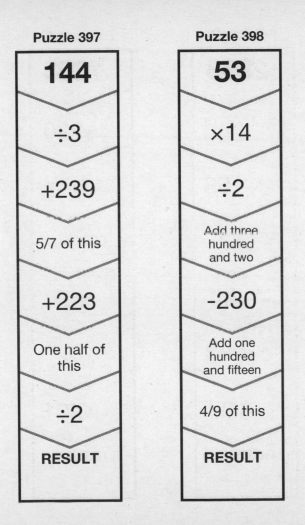

Puzzle 399

253

17/23 of this

+156

÷7

-14

+283

-109

RESULT

Puzzle 400

293

-230

5/9 of this

60% of this

+57

÷2

+421

RESULT

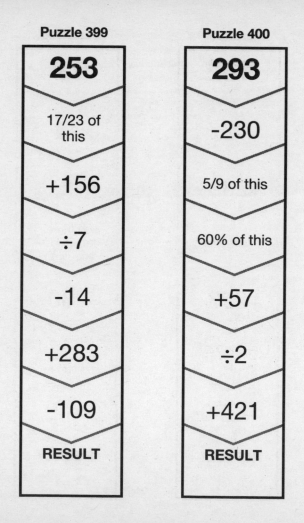

Puzzle 401

162
+351
5/9 of this
-260
×17
40% of this
+370
RESULT

Puzzle 402

78
÷2
2/3 of this
+714
15% of this
-74
×19
RESULT

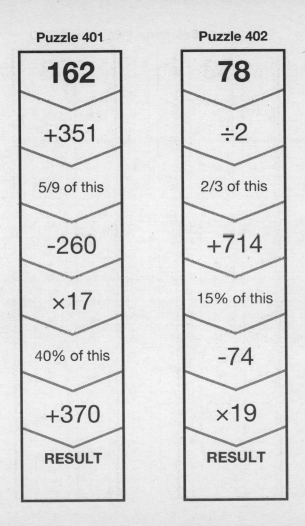

Solutions 1-24

1	2	3	4	5	6	7	8
16	**11**	**15**	**40**	**47**	**18**	**49**	**50**
8	6	5	20	36	9	7	77
3	3	50	14	6	3	26	35
21	38	25	2	30	33	13	7
7	76	23	12	10	11	11	28
77	19	36	17	54	27	37	42
65	**4**	**12**	**16**	**29**	**5**	**29**	**14**

9	10	11	12	13	14	15	16
45	**21**	**7**	**39**	**28**	**7**	**15**	**5**
15	7	28	35	14	35	41	60
5	49	14	5	11	5	4	45
55	72	15	25	15	12	2	46
11	24	5	23	5	6	22	29
6	42	38	26	37	3	11	70
60	**47**	**19**	**13**	**36**	**36**	**50**	**35**

17	18	19	20	21	22	23	24
17	**7**	**50**	**27**	**46**	**36**	**32**	**13**
20	56	10	3	23	72	4	12
10	28	5	41	41	18	2	6
5	24	45	2	10	2	14	2
27	2	36	14	5	3	7	3
9	4	57	7	51	17	56	22
3	**42**	**19**	**44**	**17**	**11**	**73**	**11**

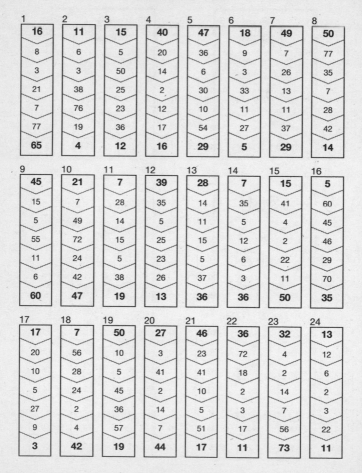

Solutions 25-48

25	26	27	28	29	30	31	32
21	**41**	**29**	**6**	**42**	**32**	**14**	**38**
7	11	21	3	14	14	7	19
56	50	42	52	28	38	49	63
28	25	51	39	49	19	72	9
4	44	17	13	7	59	12	45
30	4	4	42	43	12	3	15
15	**28**	**24**	**21**	**10**	**3**	**18**	**3**

33	34	35	36	37	38	39	40
21	**19**	**12**	**34**	**5**	**5**	**18**	**36**
3	64	6	17	50	43	27	18
25	16	53	38	23	42	9	72
5	8	40	19	56	21	20	54
20	2	48	41	42	7	5	27
4	3	44	30	14	10	34	20
2	**27**	**11**	**15**	**7**	**40**	**13**	**5**

41	42	43	44	45	46	47	48
17	**22**	**19**	**18**	**33**	**28**	**15**	**21**
10	40	4	9	3	7	5	7
5	20	32	52	21	27	25	28
25	5	24	13	7	9	22	4
16	47	12	32	56	3	11	2
8	33	48	16	19	16	39	10
55	**11**	**72**	**8**	**61**	**2**	**13**	**5**

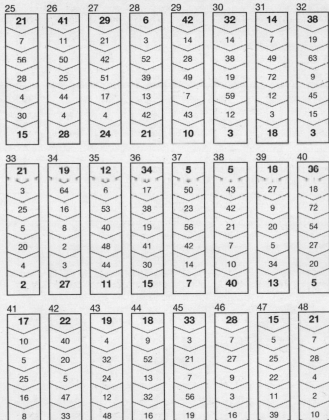

Solutions 49-72

49	50	51	52	53	54	55	56
9	**37**	**28**	**21**	**37**	**12**	**24**	**34**
3	24	70	7	4	2	30	17
33	8	10	56	8	10	5	51
11	2	40	74	80	5	10	26
5	30	43	42	54	20	11	13
25	3	22	14	18	35	77	65
73	**48**	**11**	**46**	**9**	**42**	**72**	**70**

57	58	59	60	61	62	63	64
38	**40**	**26**	**36**	**49**	**45**	**6**	**26**
19	20	13	12	60	15	2	13
57	10	61	4	12	32	23	30
72	5	41	2	56	18	22	5
9	27	80	10	8	9	11	15
3	9	40	22	2	2	5	35
6	**36**	**20**	**11**	**3**	**4**	**25**	**7**

65	66	67	68	69	70	71	72
9	**6**	**27**	**38**	**12**	**27**	**25**	**40**
3	3	9	19	24	3	5	14
30	48	3	57	6	6	60	2
36	16	38	63	3	2	20	8
20	2	19	56	42	37	34	6
5	44	11	28	32	36	33	3
38	**22**	**29**	**4**	**61**	**48**	**11**	**33**

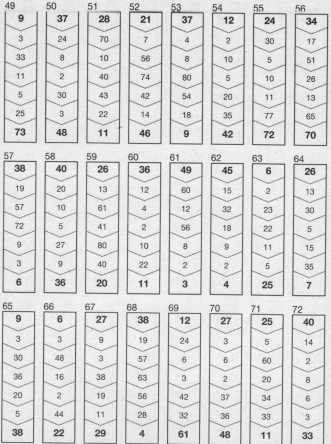

Solutions 73-96

73	74	75	76	77	78	79	80
46	**31**	**34**	**5**	**9**	**27**	**22**	**26**
51	32	17	45	18	38	11	13
17	8	8	79	36	19	7	62
68	4	20	76	6	57	77	31
34	2	10	38	66	50	48	42
71	18	5	19	33	25	72	6
63	**27**	**25**	**46**	**11**	**75**	**36**	**54**

81	82	83	84	85	86	87	88
16	**15**	**22**	**31**	**37**	**14**	**19**	**20**
8	5	11	37	66	7	38	35
72	3	74	12	33	84	57	15
36	9	37	6	11	42	31	90
29	90	19	4	22	54	66	60
31	18	58	48	39	18	29	18
9	**88**	**29**	**72**	**21**	**6**	**85**	**9**

89	90	91	92	93	94	95	96
21	**23**	**25**	**36**	**29**	**47**	**43**	**8**
14	46	15	12	78	94	86	4
7	69	9	6	26	30	15	48
35	33	3	15	97	27	29	78
28	3	29	3	47	18	4	39
68	19	24	39	94	9	32	19
22	**18**	**12**	**5**	**74**	**3**	**8**	**64**

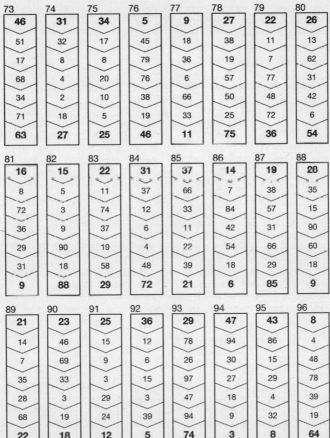

Solutions 97-120

97	98	99	100	101	102	103	104
22	**33**	**16**	**20**	**24**	**8**	**35**	**34**
45	34	8	10	12	78	7	17
12	17	20	54	96	6	20	51
6	45	14	36	48	42	5	83
50	15	25	9	51	63	4	77
48	60	5	74	17	7	28	7
24	**9**	**4**	**60**	**34**	**35**	**42**	**84**

105	106	107	108	109	110	111	112
14	**6**	**27**	**45**	**38**	**43**	**43**	**24**
45	42	1	18	19	38	91	12
27	14	49	9	36	19	7	96
9	7	7	20	18	11	15	64
3	62	76	40	54	22	10	8
33	30	38	28	24	82	90	40
60	**10**	**19**	**44**	**12**	**41**	**45**	**32**

113	114	115	116	117	118	119	120
43	**19**	**20**	**14**	**37**	**37**	**49**	**44**
98	95	10	7	74	33	10	22
47	38	80	52	40	11	17	59
77	61	32	49	16	45	82	57
4	36	59	59	4	5	72	38
32	54	45	50	32	50	12	19
48	**27**	**27**	**30**	**27**	**69**	**13**	**77**

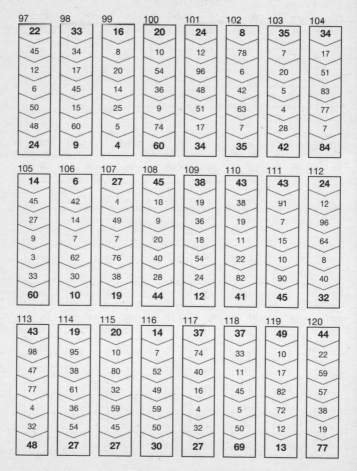

Solutions 121-144

121	122	123	124	125	126	127	128
18	**50**	**25**	**42**	**46**	**34**	**49**	**10**
9	25	5	21	23	17	57	5
81	5	15	3	21	44	19	24
27	55	27	12	7	35	85	4
54	16	9	6	14	7	48	19
57	8	36	90	84	77	24	95
19	**4**	**6**	**59**	**42**	**92**	**3**	**22**

129	130	131	132	133	134	135	136
33	**46**	**24**	**36**	**29**	**26**	**40**	**45**
66	23	8	12	61	39	16	9
6	69	11	6	40	13	41	54
48	77	55	3	8	11	7	46
24	7	31	70	48	66	63	23
36	25	74	10	12	30	9	95
81	**10**	**37**	**62**	**57**	**27**	**23**	**38**

137	138	139	140	141	142	143	144
34	**22**	**22**	**14**	**9**	**41**	**38**	**10**
17	33	35	7	3	40	19	81
88	76	21	70	64	56	34	27
51	38	64	10	32	74	17	3
64	19	11	5	16	37	49	43
8	4	99	4	47	5	7	8
4	**24**	**33**	**32**	**19**	**20**	**56**	**4**

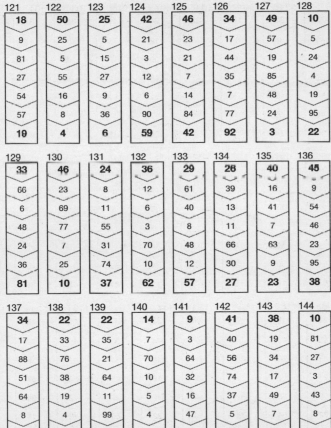

Solutions 145-168

145	146	147	148	149	150	151	152
18	**40**	**35**	**25**	**15**	**24**	**41**	**26**
23	8	5	20	6	3	74	13
69	4	15	55	3	71	39	42
83	52	18	5	12	11	60	21
55	26	9	35	71	44	18	7
5	13	3	11	62	22	9	35
10	**49**	**10**	**68**	**31**	**96**	**3**	**14**

153	154	155	156	157	158	159	160
40	**50**	**32**	**19**	**15**	**25**	**19**	**31**
15	30	16	70	12	5	32	22
9	4	4	38	63	10	16	11
3	52	49	4	26	51	75	33
30	26	27	29	13	17	45	3
81	78	87	22	32	28	5	44
25	**39**	**19**	**11**	**64**	**3**	**25**	**66**

161	162	163	164	165	166	167	168
28	**34**	**21**	**81**	**17**	**14**	**38**	**68**
7	76	14	63	72	7	48	34
76	38	98	12	36	15	22	51
72	19	49	36	18	5	132	136
36	16	7	6	6	47	66	59
6	69	59	30	9	188	105	119
90	**22**	**118**	**15**	**66**	**141**	**15**	**17**

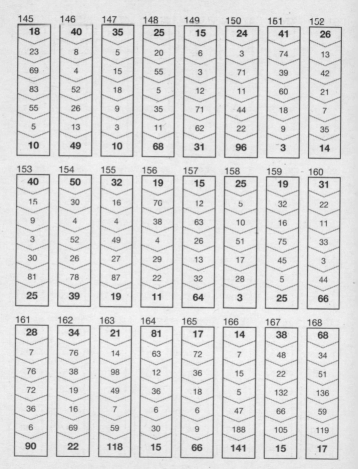

Solutions 169-192

169	170	171	172	173	174	175	176
64	**24**	**21**	**87**	**59**	**94**	**6**	**90**
48	59	7	29	92	47	3	180
145	34	62	52	23	51	36	165
29	17	31	26	103	17	4	55
116	107	56	11	76	83	64	42
81	52	4	70	152	75	163	21
150	**26**	**48**	**28**	**126**	**60**	**161**	**86**

177	178	179	180	181	182	183	184
75	**6**	**60**	**26**	**46**	**12**	**30**	**50**
45	90	9	13	50	4	112	25
120	45	3	90	35	70	48	20
24	9	54	72	7	21	46	120
6	3	27	59	105	3	23	187
3	68	135	66	63	18	17	17
33	**53**	**108**	**33**	**44**	**19**	**153**	**88**

185	186	187	188	189	190	191	192
91	**50**	**38**	**58**	**93**	**31**	**85**	**39**
138	20	15	29	31	27	34	73
118	18	75	6	11	9	68	62
59	36	46	12	36	3	74	93
61	6	23	3	6	17	37	184
183	3	138	48	3	5	117	46
123	**15**	**194**	**24**	**48**	**65**	**91**	**23**

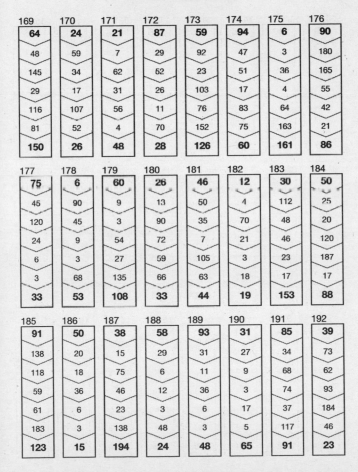

Solutions 193-216

193	194	195	196	197	198	199	200
66	**75**	**74**	**88**	**30**	**90**	**89**	**11**
105	30	37	11	180	30	178	121
35	20	148	165	126	3	85	123
14	5	54	15	63	12	34	82
112	71	18	9	17	9	136	41
16	33	126	108	136	144	81	126
83	**39**	**183**	**54**	**95**	**145**	**54**	**63**

201	202	203	204	205	206	207	208
22	**80**	**5**	**82**	**28**	**84**	**59**	**19**
11	125	65	41	7	14	31	105
52	25	78	99	51	140	61	45
26	3	43	33	17	183	15	9
78	12	172	22	14	61	3	6
136	6	157	11	105	59	45	3
85	**45**	**199**	**143**	**35**	**141**	**30**	**36**

209	210	211	212	213	214	215	216
7	**47**	**28**	**54**	**66**	**45**	**84**	**59**
6	46	14	18	6	15	28	58
60	23	24	6	3	50	56	87
33	34	36	89	21	27	48	67
11	17	18	63	114	70	74	159
90	153	3	105	76	42	37	53
72	**175**	**62**	**21**	**104**	**21**	**85**	**93**

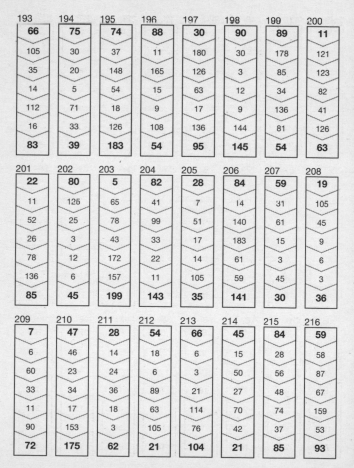

Solutions 217-240

217	218	219	220	221	222	223	224
22	**42**	**16**	**28**	**26**	**61**	**51**	**38**
11	21	111	56	39	12	17	17
132	5	37	8	43	48	53	68
66	15	5	55	6	24	21	44
106	25	58	38	8	6	63	94
53	125	87	91	80	90	9	47
26	**50**	**20**	**52**	**49**	**45**	**66**	**75**

225	226	227	228	229	230	231	232
18	**55**	**84**	**88**	**72**	**59**	**55**	**63**
9	5	42	22	18	26	33	21
76	3	14	11	103	13	3	7
19	10	109	187	65	88	60	91
171	50	89	154	52	18	57	78
81	60	164	77	26	6	38	39
164	**17**	**110**	**101**	**182**	**24**	**56**	**76**

233	234	235	236	237	238	239	240
54	**51**	**44**	**69**	**83**	**65**	**33**	**64**
15	153	22	23	11	52	67	8
39	190	11	161	65	22	22	4
13	168	61	182	39	11	11	3
91	126	26	91	13	100	132	20
167	63	13	13	117	56	66	5
78	**21**	**50**	**35**	**64**	**42**	**35**	**30**

Solutions 241-264

241	242	243	244	245	246	247	248
48	**88**	**110**	**124**	**32**	**40**	**58**	**129**
36	22	178	62	16	229	29	141
135	37	89	31	176	99	170	47
108	20	31	27	22	33	85	44
183	100	186	162	118	195	184	132
130	50	62	18	177	194	92	226
91	**122**	**69**	**234**	**28**	**97**	**62**	**197**

249	250	251	252	253	254	255	256
40	**145**	**130**	**30**	**139**	**74**	**22**	**80**
15	29	65	25	131	37	33	204
155	48	25	67	234	171	109	153
130	32	40	64	117	19	69	34
65	16	99	16	80	38	23	17
39	240	63	160	189	39	158	193
237	**156**	**51**	**235**	**63**	**33**	**79**	**90**

257	258	259	260	261	262	263	264
17	**52**	**83**	**53**	**62**	**93**	**141**	**51**
170	26	166	91	31	62	47	17
68	130	187	182	59	31	140	145
34	250	120	26	118	217	66	95
118	175	84	39	177	166	96	227
59	25	138	70	41	83	48	172
159	**35**	**69**	**62**	**228**	**232**	**144**	**86**

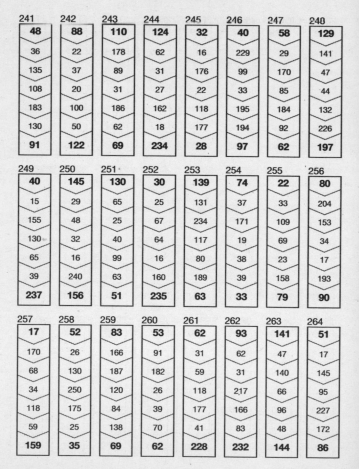

Solutions 265-288

265	266	267	268	269	270	271	272
74	**129**	**16**	**69**	**35**	**85**	**57**	**83**
118	43	20	107	210	105	19	246
85	140	46	214	168	21	47	82
68	175	23	94	112	15	235	41
17	70	184	47	159	170	94	110
136	80	161	197	139	134	247	55
42	**240**	**162**	**195**	**197**	**67**	**38**	**220**

273	274	275	276	277	278	279	280
107	**117**	**16**	**130**	**55**	**74**	**16**	**95**
19	66	128	26	128	37	192	191
216	33	160	141	32	82	96	186
72	68	60	47	16	41	24	93
54	51	99	235	144	23	18	31
59	140	85	71	243	115	150	59
177	**93**	**119**	**105**	**189**	**67**	**30**	**35**

281	282	283	284	285	286	287	288
69	**26**	**129**	**90**	**52**	**95**	**125**	**100**
46	78	86	40	26	57	150	25
159	39	43	31	130	246	25	218
53	133	62	100	115	164	134	109
111	35	38	75	46	41	67	21
37	230	106	15	62	179	124	126
35	**138**	**53**	**180**	**31**	**108**	**31**	**105**

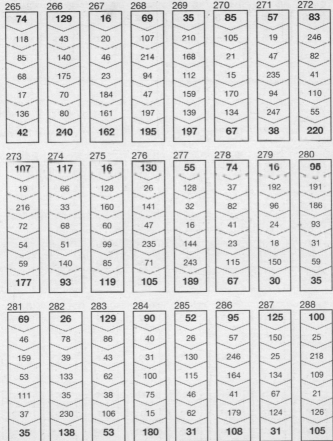

Solutions 289-312

289	290	291	292	293	294	295	296
110	**140**	**70**	**40**	**30**	**49**	**125**	**41**
77	70	35	20	18	21	75	205
184	168	184	19	72	126	240	82
92	55	46	37	54	42	30	241
23	44	110	32	248	28	15	139
218	138	30	16	56	156	117	157
109	**46**	**54**	**224**	**28**	**127**	**78**	**151**

297	298	299	300	301	302	303	304
49	**112**	**48**	**15**	**121**	**80**	**94**	**53**
147	250	16	161	99	182	47	41
63	76	169	121	33	128	74	84
113	84	52	66	224	214	37	42
98	42	26	172	64	107	148	21
84	36	171	86	104	104	111	17
125	**27**	**57**	**43**	**52**	**52**	**34**	**87**

305	306	307	308	309	310	311	312
92	**56**	**69**	**97**	**57**	**125**	**22**	**131**
161	28	46	51	19	25	220	50
23	66	230	156	34	150	20	250
32	33	115	52	17	115	176	150
16	17	169	26	136	46	87	25
103	107	140	141	77	81	145	58
206	**214**	**133**	**47**	**234**	**27**	**58**	**29**

Solutions 313-336

313	314	315	316	317	318	319	320
98	**117**	**80**	**150**	**35**	**105**	**24**	**123**
49	234	30	50	175	35	18	246
88	26	229	20	105	181	72	82
44	221	54	78	42	78	36	127
172	153	27	52	28	39	201	29
86	98	165	26	112	101	65	87
31	**141**	**150**	**134**	**108**	**202**	**104**	**158**

321	322	323	324	325	326	327	328
162	**343**	**347**	**389**	**445**	**148**	**450**	**396**
81	147	579	60	89	156	321	36
124	376	386	72	74	39	616	252
62	564	288	32	207	44	385	39
93	282	144	576	69	22	308	159
31	409	24	24	263	110	283	53
287	**287**	**303**	**299**	**225**	**104**	**560**	**516**

329	330	331	332	333	334	335	336
228	**112**	**387**	**86**	**437**	**183**	**306**	**312**
114	474	129	56	104	61	612	156
516	237	86	28	208	732	357	234
129	158	50	333	52	669	744	283
223	244	364	111	32	223	248	566
669	61	52	321	480	191	124	373
595	**610**	**102**	**214**	**688**	**377**	**496**	**684**

Solutions 337-360

337	338	339	340	341	342	343	344
215	**488**	**93**	**156**	**148**	**238**	**350**	**208**
288	61	61	78	74	119	304	104
96	307	88	39	37	277	76	502
24	52	44	462	366	158	468	251
408	26	22	32	244	389	390	30
233	392	428	498	404	206	38	21
458	**196**	**214**	**166**	**202**	**103**	**249**	**252**

345	346	347	348	349	350	351	352
100	**498**	**318**	**490**	**170**	**322**	**358**	**412**
98	166	176	245	102	40	35	103
49	83	402	49	51	225	595	348
355	245	134	222	478	50	238	29
71	231	67	333	239	700	678	377
710	299	275	113	41	140	565	614
346	**276**	**550**	**452**	**227**	**275**	**204**	**307**

353	354	355	356	357	358	359	360
77	**122**	**424**	**90**	**424**	**75**	**341**	**382**
22	61	212	162	212	50	186	90
66	260	53	152	53	44	31	196
99	142	28	617	106	264	558	98
464	71	475	572	109	132	62	49
290	497	380	286	654	66	265	441
255	**284**	**260**	**517**	**590**	**330**	**246**	**147**

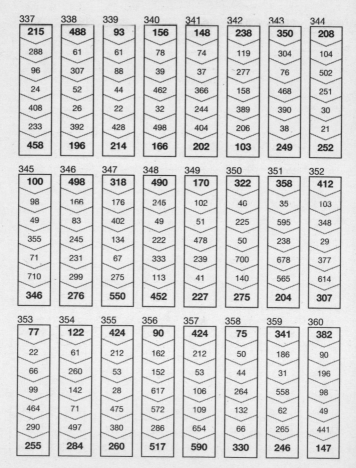

Solutions 361-384

361	362	363	364	365	366	367	368
177	**337**	**121**	**298**	**179**	**345**	**165**	**298**
59	186	341	149	95	69	77	149
472	62	155	516	38	304	294	745
118	441	31	43	500	152	441	596
323	21	340	91	125	76	691	669
76	546	216	26	100	137	265	223
117	**112**	**108**	**598**	**532**	**411**	**106**	**205**

369	370	371	372	373	374	375	376
182	**495**	**483**	**166**	**367**	**319**	**303**	**255**
104	198	335	83	565	87	532	571
26	253	268	160	452	180	377	198
413	23	134	136	113	72	29	132
59	437	356	68	331	495	442	66
478	184	89	229	204	33	221	22
448	**552**	**368**	**120**	**136**	**330**	**169**	**511**

377	378	379	380	381	382	383	384
156	**192**	**302**	**382**	**150**	**127**	**187**	**259**
52	371	151	295	190	488	226	26
520	53	209	59	95	416	113	179
208	265	132	124	43	104	266	102
104	159	66	62	458	52	86	34
728	136	33	557	254	73	43	36
610	**574**	**435**	**455**	**127**	**219**	**645**	**468**

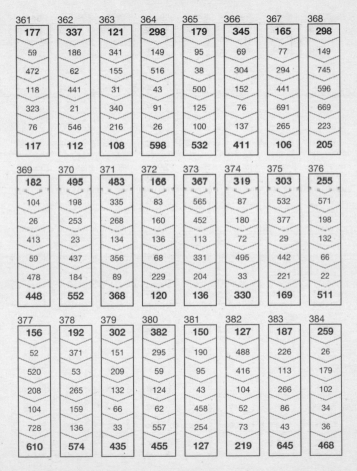

Solutions 385-402

385	386	387	388	389	390	391	392
273	**110**	**305**	**233**	**151**	**62**	**167**	**224**
168	489	292	666	604	120	95	281
84	163	146	290	302	24	239	142
28	92	182	29	549	72	30	581
439	644	91	580	545	36	750	249
340	28	39	566	436	324	375	231
170	**308**	**116**	**283**	**219**	**108**	**737**	**154**

393	394	395	396	397	398	399	400
300	**314**	**296**	**163**	**144**	**53**	**253**	**293**
137	157	148	642	48	742	187	63
533	233	74	428	287	371	343	35
369	69	226	214	205	673	49	21
709	552	85	158	428	443	35	78
478	138	51	79	214	558	318	39
239	**207**	**131**	**166**	**107**	**248**	**209**	**460**

401	402
162	**78**
513	39
285	26
25	740
425	111
170	37
540	**703**

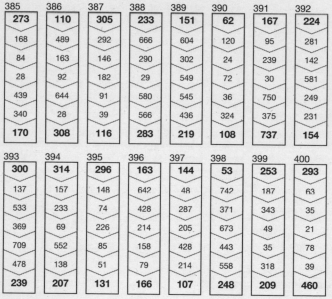